Environmental Security
in South-Eastern Europe

NATO Science for Peace and Security Series

This Series presents the results of scientific meetings supported under the NATO Programme: Science for Peace and Security (SPS).

The NATO SPS Programme supports meetings in the following Key Priority areas: (1) Defence Against Terrorism; (2) Countering other Threats to Security and (3) NATO, Partner and Mediterranean Dialogue Country Priorities. The types of meeting supported are generally "Advanced Study Institutes" and "Advanced Research Workshops". The NATO SPS Series collects together the results of these meetings. The meetings are co-organized by scientists from NATO countries and scientists from NATO's "Partner" or "Mediterranean Dialogue" countries. The observations and recommendations made at the meetings, as well as the contents of the volumes in the Series, reflect those of participants and contributors only; they should not necessarily be regarded as reflecting NATO views or policy.

Advanced Study Institutes (ASI) are high-level tutorial courses to convey the latest developments in a subject to an advanced-level audience

Advanced Research Workshops (ARW) are expert meetings where an intense but informal exchange of views at the frontiers of a subject aims at identifying directions for future action

Following a transformation of the programme in 2006 the Series has been re-named and re-organised. Recent volumes on topics not related to security, which result from meetings supported under the programme earlier, may be found in the NATO Science Series.

The Series is published by IOS Press, Amsterdam, and Springer, Dordrecht, in conjunction with the NATO Emerging Security Challenges Division.

Sub-Series

A.	Chemistry and Biology	Springer
B.	Physics and Biophysics	Springer
C.	Environmental Security	Springer
D.	Information and Communication Security	IOS Press
E.	Human and Societal Dynamics	IOS Press

http://www.nato.int/science
http://www.springer.com
http://www.iospress.nl

Series C: Environmental Security

Environmental Security in South-Eastern Europe

International Agreements and Their Implementation

edited by

Massimiliano Montini
University of Siena
Department of Economic Law
Siena, Italy

and

Slavko Bogdanovic
University Business Academy
Novi Sad, Serbia

Published in Cooperation with NATO Emerging Security Challenges Division

Proceedings of the NATO Advanced Research Workshop on
International Regulation Implementation for Environmental Security
in South Eastern Europe Venice, Italy
3–4 December 2009

ISBN 978-94-007-0245-5 (PB)
ISBN 978-94-007-0227-1 (HB)
ISBN 978-94-007-0228-8 (e-book)

Published by Springer,
P.O. Box 17, 3300 AA Dordrecht, The Netherlands.

www.springer.com

Printed on acid-free paper

All Rights Reserved
© Springer Science + Business Media B.V. 2011
No part of this work may be reproduced, stored in a retrieval system, or transmitted in any form-or by any means, electronic, mechanical, photocopying, microfilming, recording or otherwise, without written permission from the Publisher, with the exception of any material supplied specifically for the purpose of being entered and executed on a computer system, for exclusive use by the purchaser of the work.

Contents

Foreword	vii
Acknowledgments	xv
Contributors	xvii

1 The Role of International Organisations in Environmental Security Issues 1
Fausto Pedrazzini

2 The Environment and Security Initiative in South Eastern Europe: Transforming Risk into Cooperation 17
Pier Carlo Sandei

3 The Role of UNESCO Designated Sites in Fostering International Cooperation and Environmental Security in SEE 27
Giorgio Andrian and Karl-Heinz Gaudry

4 Social and Environmental Issues Related to Security in SEE Countries 43
Dragoljub Todić

5 The Impact of International Treaties on Climate Change in SEE Countries 59
Massimiliano Montini

6 The Impact of the International Treaties on Water Management in South-Eastern Europe 77
Slavko Bogdanovic

7   Towards Environmental Security by Adapting
    the Energy Sector: Summary of Strategies and Opportunities
    for Technology Transfer and Cooperation in the SEE Region 95
    Vanessa Peña and Brandon Petelin

8   Challenges of Environmental Protection in Times
    of Armed Conflict.. 119
    Robert Mrljić

9   Food Security and Eco-terrorism Impacts on Environmental
    Security Through Vulnerabilities 137
    Hami Alpas and Taylan Kiymaz

10 Emergency Situations and Risk Management
    in Multilateral Environmental Agreements:
    A Comparative Analysis in the SEE... 151
    Gordana Petkovic

11 Large Scale Infrastructural Projects in South Eastern Europe
    and Their Impact on Political Relations, Economic
    Development and Environmental Security 165
    Dimitar Pekhlivanov

12 Environmental Impact Assessment in a Trans-boundary
    Context in the SEE Countries 183
    Nataša Đereg

13 Promoting Environmental Protection Through the
    Management of Shared Natural Resources Between
    Albania and Montenegro: The Shkodra Lake Watershed 197
    Djana Bejko

14 Environmental Impact Assessment in a Trans-Boundary
    Context in Montenegro 213
    Lazarela Kalezić

15 The Framework Agreement on the Sava River Basin (FASRB) 229
    Jasnica Klara Matic

Foreword

I

Nowadays, many environmental issues have security implications, above all in vulnerable regions that face multiple stresses, such as past conflict effects, internal and trans-boundary political instability, weak institutions, poverty and unequal access to resources, food insecurity and so on. Climate change and water management appear to be two of the main relevant issues contributing to the increase of instability risks, in particular in those countries where conflicts and tensions are already present and a greater deal of adaptation efforts are required due to their special environmental conditions. In this sense, it is well known that South-Eastern Europe (SEE) is particularly at risk as concerns water scarcity, desertification, changes in soil humidity and suitability to traditional agriculture and deriving possible conflicts.

Cooperation among different sovereign states is essential to better tackle these cross-border challenges and international environmental agreements may play an important role, as tools which contribute to control and to reduce negative impacts. The NATO Advanced Research Workshop (ARW) titled "International Regulation Implementation for Environmental Security in South-Eastern Europe", which was held at the Venice International University (Italy) in December 2010 explored these topics with the final goal of promoting security and international relations between the SEE countries sharing environmental resources.

The Workshop was coordinated by Massimiliano Montini (University of Siena, Italy) and Slavko Bogdanovic (Business Academy, Novi Sad, Serbia), in co-operation with Ilda Mannino (Venice International University, Italy), and convened 21 experts from academia, research institutions, international organizations, government and NGOs, with different specialization in the field of environmental legislation and security, coming from Albania, Austria, Bosnia Herzegovina, Bulgaria, Croatia, the Former Yugoslav Republic of Macedonia (hereinafter Macedonia),[1] Hungary, Italy, Montenegro, Romania, Serbia, Slovenia, Turkey and the USA.

The specific aims of the Workshop were to highlight the main environmental issues related to security in the SEE countries, with a particular focus on climate change and water management, to analyse the most relevant international agreements in the above mentioned areas, to assess the major implementation challenges related to those agreements and to evaluate whether their correct implementation may contribute to reduce risks to environmental security.

II

This book gathers the main outcomes of the workshop and points out the regional and national efforts made in environmental regulation implementation, as well as the main challenges and opportunities.

Within a global perspective, Pedrazzini starts from the assumption that environmental degradation has a direct effect on the stability of populations and might trigger tensions and conflicts. For this reason, the concept of environmental security has evolved at international level as a subject matter where are involved not only the organizations dealing with environmental issues, but also those dealing with security, policy and economic issues. Within such a context, the author observes that the implementation of collaborative programmes and projects between those organisations together with the traditional environment agencies and institutions is an innovative and multidisciplinary approach which can broaden the perspective for improving the conditions of the environment and of the populations living on it.

In the same sense, co-operation is the focus of the analysis by Sandei which presents the results of the ENVSEC (Environment and Security) Initiative, a unique partnership between several UN and non-UN agencies targeting certain hotspots in Europe and Central Asia, which is delivering good results for SEE countries and could serve as a model to meet security and environmental concerns also in other areas of the world.

Andrian and Gaudry analyse the recent political changes in the SEE region, related mainly to the breaking of the Socialist Federal Republic of Yugoslavia which resulted in the creation of a number of new boundaries. The consequence of this process in many cases is the presence of a high conflict potential due to unresolved border issues. Under such conditions, the role of international organizations fostering several co-operation activities and initiatives proved to

---

[1] The Former Yugoslav Republic of Macedonia (FYROM), is registered with this name at the United Nations, but claims the possibility to use the name "Republic of Macedonia". Over the use of this name there is an unsettled dispute with the Republic of Greece. Therefore, in the present publication the names "Republic of Macedonia" or "Macedonia" will be used for mere reasons of shortness. This choice does not imply taking any position on the pending controversy over the use of the name which is to be settled under International Law.

be a powerful tool to try and reconcile those tensions. The authors illustrate such a view presenting the example of the UNESCO designated sites, namely, the *World Heritage Sites* and the *Biosphere Reserves*, characterised by their international visibility and specific governance mechanisms which guarantee a certain degree of multilateral cooperation.

Concentrating his attention on the issues connected to the relation between social and environmental issues on one side and security risks on the other side, Todić highlights that several important social issues which characterise the SEE countries (such as their geographic position, level of economic development and cooperation, historical heritage, state of awareness of the public concerning specific social questions, relation with the EU) represent the basic factors which influence the understanding of environmental security in the SEE region. Therefore, the author pledges for the implementation of the existing international (global and regional) legal framework for strengthening cooperation between the SEE countries, possibly working at the same time on the development of further specific regional multilateral environmental agreements and supporting the EU integration process under way.

In his paper concerning the impact of the international treaties on climate change, Montini elaborates on the link between climate change and environmental security, highlighting in particular the contribution that climate change may give to the worsening of the environmental crisis at a global level, acting as a "threat multiplier". Following the conclusion that a serious sharp series of climate change mitigation and adaptation activities, aimed at the prevention of the possible combined negative effect of climate change and environmental security, must be undertaken in the years to come, the author addresses more specifically the issue of how to reduce environmental security risks in SEE through the implementation of climate change treaties. In such a context, in particular, two specific scenarios for SEE countries are proposed and analysed. Firstly, under option A, the possibility to implement the existing major climate change treaties is considered. Secondly, under option B, a more sophisticated alternative is considered, based on the track indicated by the 2005 *Energy Community South East Europe Treaty* (ECSEE Treaty).

The following paper by Bogdanovic provides a comprehensive review of the existing legislative framework applicable to water resources in SEE. The author finds that the current national legal regimes on water management in the SEE countries have been greatly influenced by the recent developments in international law and by the EU water management policy. It then identifies some open fields for cooperation between the SEE countries and also certain specific issues that could cause tensions and become subjects for disputes between the SEE countries sharing the same water resources. In this sense, the author sees on the one side the acceptance of the relevant UNECE water management and environmental treaties and the full compliance with them, as well as with the EU *acquis*, and on the other side the introduction of modern water management systems capable to cope with the various challenges as the most suitable tools which may be used for avoiding water disputes in the SEE countries.

Peña and Petelin deal with the opportunities for the energy sector adaptation in response to environmental, energy and climate security concerns. They survey five fundamental criteria for an effective environmental regime within the context of renewable energy and energy efficiency policy in the SEE region. Their evaluation leads them to the conclusion that the countries in the SEE region have made significant progress in developing renewable energy and energy efficiency policies, programmes and projects, but there remain regulatory, economic, and institutional barriers to effective energy adaptation, which could be overcome through technology transfer and capacity building as well as regional collaboration initiatives.

On a different perspective, Mrljić presents a short analysis of the most relevant current international law provisions related to the protection of the natural environment in times of armed conflicts. Beside dealing with the general international law provisions in this field, the paper contains a reference to the possibilities relating to the application of peacetime environmental treaties for the times of armed conflict. Moreover, the author focuses on the rules related to the protection of water resources in times of armed conflict.

Alpas and Kiymaz focus on the link between food security and climate change. Starting from the assumption that food security is directly affected by climate change, they find that food security requires appropriate agricultural management and utilization of natural resources and eco-systems, as well as good governance and sustainable political systems. They observe that the impact of climate change in conjunction with potential food prices volatility could increase food insecurity and have severe impact on poor groups, thus leading to social disturbances and terrorism in the short to long term. In order to prevent that, the authors plead for replacing traditional agricultural policies which have a negative effect on the environment with new policy alternatives for an environmentally friendly support to food security. To this effect, they recommend to make recourse to the vulnerability assessment, as a concrete tool that could help in addressing food supply-chain security by determining the selection and ranking of countermeasures and emergency responses to the global environmental change.

Dealing with the issue of emergency response and risk management in the SEE countries, Petkovic analyses the international legal regime dealing with the prevention of trans-boundary natural and industrial disasters which was developed by the UNECE and is applicable to the SEE area. A particular attention is paid to the Framework Agreement on the Sava River Basin (FASRB) and the implementing national legislation in the SEE countries. The author finally concludes that SEE countries should strive to develop and implement prevention and risk reduction measures integrated into their sustainable development policies, including climate change as one of their key components.

Considering the obvious economic backwardness of the SEE region, Pekhlivanov elaborates on the role of EU as an "anchor" for the development of the area. Several examples of large-scale "trans-regional" or "trans-Balkan" infrastructure projects, such as gas and oil pipelines, bridges and so on, are reviewed in this paper. A specific reference is made to the environmental risks and issues that may emerge during the realization of such projects. In spite of the possible shortcomings, however, the author concludes that such projects should be welcomed insofar they could make a breakthrough in the borders of the region by linking the SEE area fast and easily with the Western Europe, fulfilling in that way the "common dream" of the local political decision-makers.

An overview of the current legal regime and the respective obligations of the SEE countries with regard to the environmental impact assessment (EIA) procedure is presented by Đereg. Following a brief analysis of the implementation of the Espoo Convention in SEE, the author analyses a new multilateral treaty signed by the SEE countries under the Espoo Convention, the so-called Bucharest Agreement. This new treaty is seen as a potentially diverting instrument, which could jeopardise the principles of the Espoo Convention, due to the special regime regarding the realization of joint trans-boundary projects established therein. In this sense, the author warns about the possible risks and proposes some suggestions to address the loopholes of the Bucharest Agreement.

As an example of a trans-boundary and shared lake watershed management and protection, Bejko presents and evaluates the existing freshwater concerns and the co-operation model developed for Shkodra/Skadar lake ecosystem, addressing the environmental security concern of the local population living in both sides (Albania and Montenegro). In such a context, the author argues that an ecosystem oriented approach is an effective tool for a correct prioritization of freshwater utilization, since it fosters cooperation and enhances a continuum from dialogue and partnership, to sharing of information, to more defined frameworks of cooperation at the trans-boundary level, to binding legal norms.

Similarly, within a trans-boundary context, Kalezić examines the national legislation of Montenegro on the environmental impact assessment (EIA) procedure, which gives full effect to the relevant international and EU legal provisions regulating the trans-boundary impacts assessment of projects. On the basis of this general and preliminary analysis, the author examines three specific cases regarding the application of the EIA procedure in a trans-boundary context in Montenegro. Such cases relate to the project for the construction of the hydroelectric power plant (HPP) "Buk-Bijela", to the HPP project "Ašta" and to the current and planned activities regarding the multi-purpose HPP project on the Morača River.

Finally, a positive example of a potentially comprehensive river basin cooperation in the post-war era in the Balkans is presented by Klara Matic. It relates to the cooperation between countries riparian to the Sava River, namely Slovenia, Croatia, BiH and Serbia, which finds its legal basis in the Framework Agreement on the Sava River Basin (FASRB). This paper contains details on the FASRB and on the Sava Commission, the body in charge for its implementation, as well as on the scope of this unique cooperation treaty referring to both navigable and non-navigable water uses, and incorporating specific management and environmental protection rules.

III

In general terms, it emerges from the discussion at the ARW and from the papers contained in the present book that there are many international environmental agreements, which cover all the wide spectrum of environmental related issues. In most cases, however such agreements lack a regional dimension and are not the most useful tools to effectively address certain environmental questions specifically related, for instance, to the SEE countries. However, the major shortcoming of the present system lies in the fact that most international environmental agreements are poorly implemented by the Parties and quite often efficient compliance regimes or dispute-settlement procedures are missing. On the basis of such premises, the focus of the ARW has been mainly on the identification of the major environmental issues, related to the implementation of the most relevant applicable international environmental agreements, which may negatively affect environmental security in the SEE region.

In the climate change sector, for instance, building on the assumption that climate change may have severe consequences for environmental security, insofar it may contribute to the worsening of the already occurring environmental crisis at a global level, acting as a "threat multiplier", the analysis has focused specifically the issue of how to reduce environmental security risks in SEE through the implementation of the major climate change treaties.

In the water management sector, the specific impact of the UNECE environmental and water treaties was unanimously recognized by the ARW participants. Besides, the specific case of the UN Convention on the Law on Non-Navigational Uses of International Watercourses was considered and it was unanimously held that all the SEE countries should ratify it.

In the same field, it was observed that there are several positive examples of multilateral and bilateral cooperation regarding watersheds and river basins. In this sense, one may refer for instance to the Framework Agreement for the Sava River Basin as well as to various initiatives developed in the forms of policy and soft law instruments, which connect the SEE region to a broader geographical context (e.g. Black Sea, Mediterranean, Central and Eastern Europe). There is a serious room for more new treaties of this category, the conclusion of which should be encouraged and supported.

The only risk to be avoided lies in the development of new bilateral or multilateral treaties which could jeopardise the already existing legal framework in certain specific sectors. A warning in this sense comes form the case of the recent Multilateral Agreement among the Countries of South-Eastern Europe for Implementation of the Convention on Environmental Impact Assessment in a Transboundary Context (so called Bucharest Treaty or "Small Espoo" Treaty), which enables SEE countries to partially divert form the consolidated Espoo Convention rules when planning and implementing trans-boundary projects. Such a treaty, in fact, could have the negative effect of jeopardising the well established legal regimes concerning EIA (at national, EU and even UNECE levels).

More generally, it should be underlined that the threat to (environmental) security in the SEE area, rooted in the unilateral/multilateral development of projects affecting trans-boundary natural resources could be high in forthcoming times, also due to the unpredicted consequences which may hit the area because of climate change and energy deficiencies. These factors increase the already existing risks associated with the traditional potential sources of disputes (e.g. unresolved State's borders, possibility of terrorist attacks on food resources-crop production and so on), thus deserving a continuing attention, particularly in a SEE countries environmental security perspective.

Finally, the work at the ARW identified new and further possibilities for cooperation between the various research institutions involved in this project. An immediate further step might lead to the development of a more structured network, aimed at developing a specific capacity for the implementation of new research projects, that would further develop and upgrade the initial findings of this ARW.

Siena—Novi Sad—Venice, 01.06.2010

The Co-editors
Massimiliano Montini
Slavko Bogdanovic

Acknowledgments

This book contains the proceedings of the Advanced Research Workshop (ARW) "International Regulation Implementation for Environmental Security in South-Eastern Europe", which was held at the Venice International University (Italy) in December 2010. The workshop was gratefully sponsored by the NATO under the Science for Peace and Security (SPS) Programme, as it is the present publication and co-funded by the Venice International University.

Therefore, we would like firstly to thank the NATO for its generous support for the workshop and the publication of the present book. The workshop and the publication would not have been possible without the scientific and technical co-operation with the TEN Center of the Venice International University (VIU). In particular, we would like to warmly thank Dr Ilda Mannino, one the main driving forces behind this project and our alter ego at VIU, as well the President Center, Prof. Ignazio Musu, and the Coordinator, Ms Alessandra Fornetti. Besides, we would like to express our sincere appreciation for the grateful support of Mr Alessandro Spezzamonte, Ms Jasmine El Din and Ms Lorenza Fasolo of VIU, who dealt with the management of the workshop administration and logistic.

Secondly, we would like to thank all the several participants to the workshop and authors of the various chapters of the book, coming from a great variety of countries, with the common aim of contributing to the advancement of research on environmental security issues in SEE.

Thirdly, a special thanks goes to the Members of the Environmental Legal Team, based at the Department of Economic Law of the University of Siena and linked to the TEN Center of VIU, who kindly helped for the production of the book, and in particular to Mr Gerardo Giannotti and Ms Francesca Volpe for the scientific support and to Mr Gabriele Bartali for the editorial support.

Finally, it should be recalled here that we decided to submit the various papers presented in this publication to a light linguistic revision, in order not to loose the cultural and linguistic variety expressed by the several authors coming from many different countries.

The Co-editors
Massimiliano Montini
Slavko Bogdanovic

Contributors

Hami Alpas
Department of Food Engineering, Middle East Technical University, Ankara, Turkey

Giorgio Andrian
UNESCO Venice Office, Regional Bureau for Science and Culture in Europe (BRESCE), Venice, Italy

Djana Bejko
Faculty of Natural Sciences, Biology Department, The Regional Environmental Center Albania and University "Luigj Gurakuqi", Shkoder, Albania

Slavko Bogdanovic
University Business Academy, Novi Sad, Serbia

Nataša Đereg
Center for Ecology and Sustainable Development, Subotica, Serbia

Karl-Heinz Gaudry
Institute for Landscape Management, University of Freiburg, Freiburg, Germany

Lazarela Kalezić
Secretariat for Spatial Planning and Environmental Protection, Podgorica, Montenegro

Taylan Kiymaz
State Planning Organization, Ankara, Turkey

Jasnica Klara Matic
EQUILIBRIUM, Environmental Law Association, Zagreb, Croatia

Massimiliano Montini
University of Siena, Department of Economic Law, Siena, Italy

Robert Mrljić
University of Zagreb, Zagreb, Croatia

Fausto Pedrazzini
CNR Institute of Clinical Physiology, Pisa, Italy

Dimitar Pekhlivanov
Institute of Public Administration, Sofia, Bulgaria

Vanessa Peña
Science and Technology Policy Institute, Institute for Defense Analyses, Washington, DC, USA

Brandon Petelin
U.S. Department of Commerce, International Trade Administration, Washington, DC, USA

Gordana Petkovic
Ministry of Environment and Spatial Planning, Belgrade, Serbia

Pier Carlo Sandei
Regional Office for Europe, UNEP, Vienna International Centre, Vienna, Austria

Dragoljub Todić
Graduate School of International Economy, Megatrend University, Belgrade, Serbia

The Role of International Organisations in Environmental Security Issues

Fausto Pedrazzini

Abstract The status of the environment is influenced by a variety of factors, the most important of which are water, land, climate and human interventions. The degradation of the environment has important consequences at all levels, including the well-being of populations. This is mainly caused by the diminution of services and resources provided by an ecosystem while it is degraded. Under these circumstances, the environmental degradation has a direct effect on the stability of populations and might trigger tensions and conflicts; this is the reason for which the concept of environmental security is becoming a subject of analysis and discussions. Such a concept goes beyond the competences of the traditional organisations and institutions which normally deal with environmental issues, and it is now taken into consideration by organisations which have their missions in other fields, like security, politics, economy and sociology. The implementation of collaborative programmes and projects between these organisations together with the traditional environmental agencies and institutions, is an innovative and multidisciplinary approach which can broaden the perspective for improving the conditions of the environment and of the populations living in it.

Keywords Environmental security • International organizations • Water management • Land management • Disaster management • Climate change • NGOs

1 Introduction

During the current and the last decades the concept of security has expanded into national, societal, human and environmental "securities", even though the concept of environmental security was officially introduced at the 42th session of the

F. Pedrazzini (✉)
CNR Institute of Clinical Physiology, Pisa, Italy
e-mail: fausto.pedrazzini@ifc.cnr.it

Schrijver (1989). However, environmental security is a continuous evolving concept; consequently an established definition does not exist (Schrijver 1989).

The relation between the environment and the security of humans has been the object of much research and the subject of many publications in recent decades, but it is only recently becoming an important focus of international environmental policy.

A stable natural ecosystem is the basic requisite for the stability and the well being of the population living on it and for its economic and social activities. The natural environment is the primary provider of essential resources like water, soil and vegetation, which are the basis for an equitable exploitation aimed at sustaining food and energy production. These resources normally exist in a natural equilibrium, which might be altered both by man made and by natural events.

The decrease of available resources, the competition to get access to them, the social and political instability and migrations, are also consequences of a degraded environment and are directly related to the security of populations (Pedrazzini 2008).

Water has long served as a key illustration of environmental security, but only in the last decades water scarcity became dramatic as a result of unprecedent growth of populations and human activities. Since 1950 the global freshwater supply per person has fallen by 60% as world population has swelled by over 150% and the world's water consumption has increased by 180%. During the past half century there have been more than 450 water-related disputes and on 37 occasions violent actions took place (Myers 2004).

What has emerged because of water deficits applies also to deforestation, land degradation, desertification and other environmental problems with widespread impact. Desertification, for instance, can generate broad scale problems for human welfare and political stability. Due to its capacity to trigger famines, poverty and international migrations, desertification is often a reason for social and political instability.

Even though environmental problems have a strong influence on the overall stability and security of populations, not all the environmental problems lead to conflict and not all the conflicts stem from environmental problems. While environmental phenomena contribute to conflicts, there are many other variables which can predispose a nation to instability and thus, in turn, make it specially susceptible to environmental problems.

National plans are the first step to counter environmental degradation, however, an increased collaboration across political boundaries is needed to cope with adverse phenomena which affect broad geographical areas. International organisations, agencies and conventions could play an important role in this respect by serving as platforms for trans-national collaboration. Such a collaboration is essential for evaluating to what extent improper water and land management affects populations in a way that reduces their security and also may lead to conflicts within the community and across political boundaries, thus affecting regional and political stability. In addition, it should be analysed in which cases the environmental degradation has shown to be a factor in security loss and by consequence a more precise definition of the concept "Environment Security" could be elaborated.

2 Water Resources Management

Water is the most precious resource of the environment, both in the quantity and quality. Worldwide, more than 430 million people currently face water scarcity and the numbers are set to rise sharply: nearly three billion people are expected to live in water-stressed countries by 2015 (Renner 2006).

Conflicting claims over water resources have been cited as a possible cause of violent clashes between nations that share rivers. However, cooperation rather than conflict has so far been the norm among riparian states (Renner 2006).

The issue of water demand has been the subject of several studies and workshops promoted by international organisations (UNEP, Blue Plan, NATO and OSCE), particularly for the Mediterranean Region and the Central Asia countries.

Water shortages are already occurring in parts of the Mediterranean region and will worsen particularly in the South and the East. The countries where the water resources are the lowest per capita will see the greatest increase in demand and face the greatest risk of their resources shrinking.

In Central Asia the socio-economic stability of the region strongly depends on water availability and land productivity. The region shelters about 50 million people, of which more than half live in rural areas and depend on irrigated agriculture. The efficient irrigated agriculture forms people's welfare, food independence and environmental security that, in turn, makes general socio-economic stability.

According to the most recent projections, a global increase of water demand is expected. To meet growing demand, national strategies essentially rely on the extension of water supply and on major waterworks to enhance resource management and reduce risks resulting from natural constraints. The supply based approach is expected to remain prevalent and lead to consequences like increased withdrawal of renewable resources; increased exploitation of non-renewable underground water resources; reuse of treated waste water for irrigation and development of interregional and international transfers. However, such an approach is reaching physical, socioeconomic and environmental limits and the factors of increasing "water vulnerability" (production costs, conflicts, sanitary risks) could be aggravated (Thibault 2009).

In order to counter the described perspective with its inevitable crises, an alternative scenario is possible (Blue Plan 2005), based on proactive policies like a more efficient an economic use of water by means of technical, economic, regulatory instruments for demand management and improved water and soil conservation.

The challenge of water demand management is not only limited to physical savings; it also means improved economic and social enhancement of mobilised resources and the coverage of water requirements of ecosystems (Thibault 2009).

3 Land/Soil Management and Desertification

An equilibrate availability of water in the ecosystem is a pre-requisite for land quality and productivity. Unfortunately a terrestrial ecosystem could easily be altered by several factors, first of all the deterioration of the soil due to different degradation processes such as erosion, loss of nutrients, compactation, salinisation and so on. The effects of soil loss or the loss of its productive and ecological functions is transferred to the other components of the ecosystem in a self-feed spiral that in its last instances gives rise to a sterile and desolate landscape (Rubio 2009).

Land degradation originates from several factors, including inaccurate agricultural practices and defective water management which cause deep alteration of the ecosystem. In some particular areas (the Mediterranean Basin, the Middle East and Central Asia), defined as arid or sub-arid, those conditions could lead to the process of desertification.

The concept of desertification has been widely discussed and analysed by specialists, but the most precise definition is the one proposed by the UN Convention to Combat Desertification (UNCCD), which states that desertification is the soil degradation in arid areas which implies the reduction or the loss of the biological productivity in agricultural soils pastures and forests (de Kabermatten 2009). Such a concept has been strengthened by Adeel et al. (2005), according to whom what characterises the phenomenon of desertification is its persistence.

Desertification means a dramatic reduction of resources and by consequence an increased competition for the residual resources. Such a situation directly affects the economic, political and social conditions of the populations living in a degraded environment.

In some regions of the globe, besides a fragile environment and limitation of available resources, political and cultural differences are also taking place, increasing the potential for tensions and conflicts from one hand and limiting the opportunity for collaborating for a better environment on the other. This is the main reason for which desertification is becoming a security issue and more specifically an Environmental security issue.

Land degradation and reduction of the biological fertility of soils, imply conditions of malnutrition for populations who become more exposed to diseases. When health and food are in danger, populations tend to leave their homes for mass migrations, frequently across political and cultural boundaries. It has been evaluated by a report of Brauch et al. (2003) that even though only 28 out of 485 conflicts which took place in the Mediterranean Region in the period 1975–2001 were caused by drought and consequent food shortage, the same conflicts involved 10.5 million people; about half of the 22 million people involved in conflicts originated by other reasons.

Within this scenario, international organisations could play a fundamental role in fostering trans-national collaborations which are essential to evaluate how the mismanagement of land and agricultural resources could strongly affect the life conditions, the stability and ultimately the security of the populations concerned.

This is the reason for which several international organisations (UN, UNEP, BLUE PLAN, OSCE and NATO) have focused their attention and organised events aimed at identifying the security aspects related to environment degradation, particularly in the Mediterranean Region.

4 Disasters Management

The general degradation of the environment is also one of the reasons (if not the most important) for natural disasters, which are indirectly induced by men. A degraded ecosystem means less protection for the populations against extreme climate conditions and consequent adverse and devastating weather phenomena.

The number of disasters has globally risen from about 750 in 1980–84 to almost 2,000 in the 2000–2004. The number of people affected has risen from about 500 million to 1.4 billion during the same period of time, and the pace is likely to accelerate in coming years as climate change translates into more intense storms, flooding, heat waves and droughts (Renner 2006).

Disasters have a devastating effect on the concerned populations in terms of people killed and/or injured; infrastructures destroyed; increased marginalisation of weak economies and generally worse living conditions of poor people and ethnic minorities: again a situation of increased non-security.

Natural disasters (e.g. earthquakes; flooding; forest fires; heat waves) are normally independent from the human activities (even though the way by which the environment is managed could play an important role in reducing or in enhancing the negative effects of natural disasters). Their effects in regions with low natural resilience could cause damages which practically induce the populations to abandon those regions.

Even though disaster prevention is the preferable course, disasters sometimes could entail unexpected developments. A disaster may inflict suffering that cuts across the divides of human conflict, prompting common relief needs and making protagonists realise that reconciliation is essential for reconstruction and recovery. A prominent example is Indonesia's Aceh province after the December 2004 tsunami. It triggered a new mood of reconciliation that allowed the 29-year conflict there to be brought to an end in 2005 (Renner 2006).

Unfortunately, disasters do not automatically translate into cooperation; particularly in the case of small scale disasters, they may not generate immediately the right positive attitude to face the difficulties. Tensions among populations and among nations could become even more intense like for instance over the distribution of relief aid. Finally, in the case of some specific disasters like droughts, the competition for the remaining water resources between different communities is the immediate consequence rather than cooperation for the common interest.

The role of international organisations is particularly relevant in countering the effects of natural disasters, both by implementing actions and preventing procedures, but also by providing help and support to the populations affected.

Countries can no longer rely on purely national solutions for large-scale emergencies, particularly given the complex nature of today's threats and the unpredictable security issues. Since it was recognised that major emergencies can pose a threat to security and stability, NATO created a body dealing with Civil Emergency Planning.

While the United Nations retain the primary role in coordinating international disasters relief, NATO provides an effective forum in which the use of civilian and military assets can be exploited to achieve a specific goal.

Beginning in the 1990s NATO has been engaged in a number of crisis responses operations, particularly in the former Yugoslavia and in the Darfur Region and also during the forest fires in Greece and in Portugal. Support was provided in the aftermath of Hurricane Katrina which devastated the United States gulf coast in August 2005. A specific cooperation between NATO and Ukraine began in 1995, following heavy rains in the Kharkiv region and support was provided during subsequent flooding.

5 Climate Change

Climate change is becoming the major environmental challenge and its consequences make it also a big security challenge. At a recent UK Foreign office briefing, maps of recent areas of conflict and civil unrest were overlaid on countries hit hardest by climate change and they were almost an exact match.

Effects of climate changes are quite important, involving a foreseen reduction of arable land in less favored regions of the world; changes in the rain patterns which will result in devastating flooding in some part of the world while others will suffer from drought; progressive ice-cap melting particularly in the North Pole regions, opening for new commercial and exploitation of resources, but at the same time creating dangerous conditions for infrastructures and territorial disputes.

However, the effect of global climate change on environmental degradation is not fully understood. Climate change may adversely affect biodiversity and exacerbate desertification due to increased evapotraspiration and likely decrease in rainfall in dry-lands, but on the other hand some dry-land species can favorable respond to this increase. Therefore, although climate change may increase aridity and desertification risk in many areas, the overall effect on an ecosystem is still difficult to predict (Ecosystems and Human Well-being – Millennium Ecosystem Assessment, 2005)

A variety of scenarios could be imagined as a consequence of the increasing evidence of a global climate change. Still this phenomenon needs to be further understood and evaluated in all its consequences. The scientific community has come a long way in elaborating sophisticated models that help us to understand the real existence and the practical consequences of the climate change on environmental changes.

Some significant examples of international cooperation on this issue can already been noted, like the Global Earth Observation summit which took place in Cape Town in 2007 where ministers and officials of over 70 governments and 40 international organisations committed to working together to develop a system to monitor what is happening to the planet. The system will allow everyone to be better informed and to work on the same data, and will also give momentum to work on collective solutions.

Climate change is an issue which goes across many other primary concerns of the modern society like energy production, storage and efficient use; economy and industrial production and development; agriculture and better use of water and available resources. All these issues need a global approach and the coherent commitment of governments and international organisations.

6 A Global Policy for a Global Safe Environment

Environmental challenges are global but they affect different areas of the world in different ways and the poorer areas experience the most dramatic effects. Environmental issues and climate change will become increasingly important at the international level, since these are issues that a country or a group of countries are not going to be able to tackle by an individual approach. There is a need of global initiatives that are able to define these problems and are also able to find multilateral solutions.

Nowadays, there is a greater awareness on what is going to happen in the future. Populations in various countries and communities at large are getting very concerned about the environment and the global change. Governments are bringing the long-term implications of today's policies and actions into the political agenda for debate, discussion and choice. There is an urgent need of political leaders who can anticipate the effects of environment degradation and climate change on populations and take position and implement initiatives that are not purely a matter of political convenience, but are rather based on the assumption that what they decide to do is good for humanity and it is good for the societies that themselves are responsible for.

Environmental challenges do not take into account political borders and require a long-term perspective and the full participation of civil societies. These are conditions which normally lead to a positive process and avoid conflicts.

Dealing with environmental issues means establishing pragmatic working programmes across political differences; establishing collaborative activities among countries, governments and populations; creating common interests and identities within homogeneous areas and around shared resources (water; soil; entire ecosystems). All this together should in perspective generate a new peaceful and safe environment.

There is a large variety of linkages among environment, resources, security, conflict and so on. Some of these links have been the subject of great attention by governments, some others less. Even in the case in which environmental issues have not been directly addressed by governments, a range of relevant programmes are developing both at the national and at the international level and projects related to a more safe and less degraded environment are having a noticeable impact on improving the interplay between environment and security in many critical areas of the world.

An example of the newly needed collaboration is related to the Mediterranean experience. The Mediterranean sea is an enclosed sea and one of the world's main waterways for shipping, with one third of petroleum trade, while the great bulk of pollutants stem from land sources in the form of industrial waste, municipal sewage and agricultural residues. In 1975 the United Nations Environment Programme started on what seemed a very ambitious project. It wanted to persuade all the coastal nations to formulate a joint strategy to tackle their common problem. Eventually 17 nations sat down to formulate a plan of action and in 1980 a conference came up with a draft of treaty, which came into operation in 1982. As a result, pollution has greatly reduced and the Mediterranean is no longer dying. There could hardly have been a region in the world with greater political disparities, yet the nations of the Mediterranean basin were persuaded to rise above their individual interests in favor of the collective welfare. The Mediterranean experience now serves as a model for parallel programmes in other regional seas (Myers 2004).

7 Who Is Who (International Organisations)/What They Do/How They Interact

The definition of international security has been debated extensively by different sources after the Second World War. As from the fall of the Berlin Wall and after the collapse of Soviet Union, the discussion over security expanded to encompass a broader range of threats to peace including particularly environmental threats associated with the political implications of resource use or pollution. Even though there is not a generalised consensus, it is quite acknowledged that environmental factors play both direct and indirect roles in political disputes and violent conflicts. The UN Millennium Project did a global assessment of the definition of environmental security and created a synthesis definition:

Environmental security is environmental viability for life support, with three sub elements:

- Preventing or repairing military damage to the environment;
- Preventing or responding to environmentally caused conflicts;
- Protecting the environment due to its inherent moral value.

These elements introduce the concept of increased concern by international organisations in dealing with environmental security issues.

7.1 NATO

As outlined in the 1999 Strategic concept, NATO plays a key role in identifying and addressing current and emerging security challenges. To this respect, it recognised that the link between environment and security is rapidly becoming a major political issue for governments across the globe. Through the Science for Peace and Security Programme, NATO Nations are helping Partner and Mediterranean dialogue countries to deal with the issues of environmental security through scientific cooperation that is delivering concrete results. Environmental Security has been identified as a key priority by NATO and in 2008 a Science Security Forum was organised in Brussels in which this issue was addressed in depth by bringing together internationally-recognised experts from all over the world. Beside this, NATO organised two workshops on Environmental Security in the Mediterranean Region: one in 2003 and one in 2007 (this in cooperation with OSCE). From both these events two relevant books were published.

The NATO SPS Programme is cooperating with the OSCE, UNDP, UNEP and REC in the field of environmental security through the ENVSEC initiative.

Ongoing SPS projects that are embedded in the ENVSEC framework include:

- Environmental Security issues Arising from the Legacy of Uranium Extraction in Central Asian Countries
- Water Resources Management of Agro-Ecosystems in the South Caucasus Transboundary Regions
- Assessment of Hazardous Chemical Contamination in the Sava River Basin (Balkan Countries)
- Study of Radio-nuclides in the Belarus Sector of the Chernobyl Power Plant's Exclusion Zone

7.2 OSCE

The Organisation for Security and Co-operation in Europe is a pan-European security body with 56 participating states. The OSCE is a primary instrument for early warning conflict prevention, crisis management and post-conflict rehabilitation in its area of competence.

Activities in the "economic and environmental" dimension include the monitoring of development among the participating states with the aim of alerting them to any threat of conflict; and assisting in the creation of environmental policies and related initiatives to promote security in the OSCE region.

OSCE has organised in September 2009 in Belgrade (Serbia) the final conference of the project "Local Environmental Security", which consisted in the organisation of a series of conferences in seven municipalities of Southern Serbia, aimed at promoting the concept of Environmental Security among the local authorities and the populations of the region. On that occasion it was stated by a member of the Serbian Government that its country should urgently comply with the regulations of the European Union on environmental issues.

7.3 UNCCD

The United Nations Convention to Combat Desertification with its 192 Parties is the only universal normative instrument dealing with land, soil and related issues. In the aftermath of its adoption, a forward looking 10 years strategy puts a renewed emphasis on the problems related to land degradation, and effectively identifies "land" as the principal subject of sustainable development, as it links the halting of land degradation and sustainable land management to the amelioration of living conditions of populations and ecosystems and the provision of global benefits.

The strategy identifies a number of objectives and expected impacts which are of interest to many processes as it points out the impact of global challenges on the everyday life of billions of people, like improvement and diversification of the livelihood base; reduction of the socio-economic environmental vulnerability of populations; enhancement of land productivity and other ecosystem goods and services in affected areas and in a sustainable manner; reduction of the vulnerability of affected ecosystems to climate change; contribution to the conservation and sustainable use of biodiversity.

7.4 UNEP

The United Nations Environment Programme is the main international institution dealing with the environment protection and the sustainable development. Its main tasks can be summarised as follows:

- Implementation of studies and analyses for monitoring the environmental conditions at the local, regional and global level
- Development of tools and procedures for the preservation of natural and environmental resources
- Support to institutions aimed at properly managing the energy resources
- Transfer of technologies and know-how for a proper and sustainable environmental development
- Build-up of a common concern on environmental issues between the public authorities, the private sector and the civil society

UNEP is a fundamental partner of the ENVSEC initiative and is the main reference for many conventions and projects. Among these, the Convention for the Protection of the Marine Environment and the Coastal region of the Mediterranean, which constitute the Mediterranean Action Plan's (MAP) legal framework, which aims at reducing pollution in the Mediterranean Sea and protect and improve the marine environment in the area, thereby contributing to its sustainable development.

7.5 UNU

The United Nations University – Institute for Environment and Human Security (UNU-EHS) is part of the UN Universities system comprised in a worldwide network of research and training centers and programmes. The concept of human security puts the individual, social groups and their livelihoods at the centre of

The Role of International Organisations in Environmental Security Issues

debate, analyses and policy. It explores problems and promotes solutions to the environmental dimension of human security. The Institute aims at academic excellence in principal priorities of its programme, which are:

- Vulnerability assessment, resilience analysis, risk management and adaptation strategies within linked human-environment systems
- Internal displacement and trans-boundary migration due to environmental push-factors

In October 2008, the UNU-EHS organised an international Conference on Environment, Forced Migration & Social Vulnerability (EFMSV). The Conference spent two and half days looking into the intriguing interplay between environmental stress inducing factors, social vulnerability and migration as a response.

EFSMV was also an element of the European Commission funded project on Environmental Change and Forced Migration Scenarios; implemented by seven institutional members of the project consortium and receiving substantial assistance by the International Organisation for Migration (IOM), OSCE and many NGOs.

7.6 European Union

The EU set the highest standards in the world to face a variety of environmental issues. The present priorities are to counter the climate change, maintain the biodiversity, reduce the pollution-induced diseases and implement a sustainable management of natural resources.

- Climate change
  The EU is actively committed towards reducing the emission of gases responsible of the greenhouse effect and has taken a leading role for launching ambitious initiatives on this issue. The EU leaders have adopted a package of rules aimed at reducing at least the 20% of the emissions of greenhouse gases and upgrading up to 20% the ratio of renewable energy resources by 2020.
- Biodiversity
  The EU plans to expand the network Natura 2000, which consists of sites in which, flora, fauna and their habitat are protected.
- Environment and Health
  The EU has adopted as its main scope to set sanitary limits for polluting substances. In addition, EU member states must monitor the presence of pollutants in the environment and intervene in case their quantity is above the limits.
- Sustainable development
  This is one of the general objectives of the EU since quite a long time. The most recent strategy of the EU on this issue combines the energy policy with the concern of climate change and stresses the importance of public education, scientific research and public support for the implementation of sustainable production and consumption models.

Two bodies of the European Union, the European Commission and the European Space Agency, launched in 2001 the initiative of the Global Monitoring System, aimed at providing by 2008 the European Union with the facility to act autonomously in the field of security and environment by means of satellite remote sensing. The objective was to make a rational use of the data collected by the satellites put into orbit by the ESA. Once put into operations, the programme GMS has been renamed Kopernikus.

Information gathered by Kopernikus helps to improve the management of natural resources, to monitor the air and water quality, to optimise agricultural activities and to promote renewable energy. Furthermore Kopernikus will enhance the safety of populations by providing early warning of natural disasters and environment degradation. In addition, it will be an essential tool for enhancing modeling activities and for better understanding the drivers of climate change.

7.7 OECD

The Organisation for Economic Co-operation and Development is an international institution which brings together 30 countries and is aimed at supporting sustainable economic growth, boosting employment, raising living standards, maintaining financial stability, assisting other countries' economic development and contributing to growth in world trade.

While the OECD does not provide direct support for the implementation of action programmes or projects to combat environmental degradation, it should be noted that its Development Assistance Committee worked on formulating policy guidance on how to mainstream global environmental issues, notably desertification in development co-operation. In addition, the OECD Environment Directorate deals with a variety of environment-related topics, such as climate change, energy and transport; consumption, production and the environment; environment in emerging and transition economies; environmental indicators and outlooks; environmental policies and instruments etc.

In a few words, the OECD is the key institution dealing with the environment and the economy and is also the leading competent international organisation dealing with the issue of subsides and sustainable development. A series of workshop was organised on this. More specifically on the occasion held in Helsinki on Subsidy Reform and Sustainable Development, the focus was on environmentally benign subsides and environmentally harmful subsides. The logic conclusion was that the latest should be removed while the benign ones should be given. An important topic like the heavy subsidisation of bio-fuels by the rich countries was also tackled. By subsidising the bio-fuels they are more competitive vis-a-vis food production, but causing adverse effect on world food situation and the alimentation of the poor. Subsidy withdrawals by the rich countries may benefit poor countries making their exports more competitive (Kiss 2006).

All these initiatives are implemented in collaboration with other international organisations like UNEP, UNCCD, EU and so on.

7.8 Think Tanks and NGOs

There is large number of Think Tanks and NGOs involved in environmental issues, programmes and projects. Their role is fundamental in increasing the public awareness of the environment and its quality. In addition, NGOs are frequently the practical executors of environment-related projects in less favored countries.

To make a detailed inventory of all these organisations would be almost impossible. Therefore, it is presented below a brief overview only about the organisations explicitly mentioned in this paper.

7.8.1 Institute for Environmental Security (IES)

IES is an international non-profit non-governmental organisation established in 2002 in The Hague, in order to increase political attention to environmental security as a means to help safeguard essential conditions for peace and sustainable development. IES has published very interesting reports, like the Introduction to the Concepts of Environmental Security and Environment Conflict, and is involved in several projects like the Climate Change & the Military Project, which is based on the cooperation of a group of leading think tanks to elaborate a message from the security sector to the December 2009 climate change negotiations taking place at the UN Conference on Climate Change in Copenhagen. It addition IES is carrying out the Enviro-Security Assessments in different areas of the world (Mataven forest in Colombia, the African Great Lakes region and Central Kalimatan – Indonesia, the Congo Basin and the Horn of Africa).

7.8.2 International Institute for Sustainable Development (IISD)

IISD is registered as a charity in Canada and is a non-profit non-governmental research institute aimed at developing expert analysis and policy recommendations to provide the knowledge needed by all citizens of the world working towards sustainable development. IISD receives financial support from Canadian federal and provincial governments, UN agencies, foundations and the private sector. The key message which is the basis of the activity of IISD is the following: Environmental degradation and the mismanagement of natural resources can fuel conflicts between and within states, contribute to poverty and state failure, and leave communities more vulnerable to disasters. At the same time environmental issues can provide an opportunity to foster dialogue and cooperation over shared interests, reducing tensions between conflicting parties.

IISD has worked with other international organisations (WWF; Wildlife Conservation Society) in central Africa to better understand the context in which these organisations operate and apply a conflict lens to their work. This work led to the development of the "Conflict-Sensitive Conservation Practitioner's Manual", which provides a decision-making framework to help conservationists understand and address natural resources-based conflicts, and integrate this understanding into conservation programming and implementation, with the scope to avoid exacerbating conflicts and maximise opportunities for peace building.

7.8.3 World Resources Institute (WRI)

WRI is an environmental think tank based in Washington (USA) and in Beijng (China), which aims at going beyond research and find practical ways to protect the Earth and improve people's lifes. It was launched in 1982 as a centre for policy research and analysis addressed to global resources and environmental issues. WRI organises its work around four key programmatic goals:

1. Climate protection
2. Governance
3. Markets and enterprises
4. People and ecosystems

WRI contributed to the publication of the report "Ecosystems and Human Well-Being; Desertification Synthesis" in 2005. This report is part of the Millennium Ecosystem Assessment (MA), which is a programme initiated in 2001 by the UN Secretary General, with the scope of assessing the consequences of ecosystem change for human well-being and the scientific basis for action needed to enhance the conservation and sustainable use of those systems. The MA has involved the work of more than 1,360 experts worldwide. Their findings, contained in five technical volumes and six synthesis report like the mentioned one on desertification, provide a state-of-the-art scientific appraisal of the conditions and trends in the world's ecosystems and the services they provide (such as clean water, food, forest products, flood control and natural resources) and the options to restore, protect or enhance their sustainable use.

8 Conclusion

By completing this paper, I realised that while dealing with the multiple components and aspect of the environment, it was immediately necessary to refer to a scenario which is the consequence of a chain of interlinked factors, including land, water, climate, human activities and so on. At the same time many other components play a role and a function in the natural environment: the biological component; the geological background and the cultural and social history also contribute to shape an ecosystem wherever it may be located in the world. Consequently, while mentioning the Environment one should be conscious of the complexity of such a concept.

Quality, preservation, sustainable, degradation, restoration, safe, polluted, equilibrate are just examples of the variety of the adjectives which normally go along with the substantive environment; security could be considered as one of the most recent ones. Security is a quite popular term during these unfortunate period of terrorism, but in the case of the environment, security identifies a concern which refers to its overall function and services towards humans. The practical effects of a non-secure environment might not be of immediate perception, still they are becoming important for the society and they are taking a high rank in the agenda of governments and international organisations.

At all levels it is recognised that environmental issues do no respect political and/or cultural boundaries, but they basically follow the general rules of the water cycle, of the meteorological events, of the natural phenomena, of the external interventions and eventually of the combination of all these factors.

The paradox is that even though the environment does not imply political boundaries, the specific regulations and interventions which might affect it are at the first approach matter of responsibility of single nations, regardless their size and political importance. This explains the reason for which the environment is essentially a resource which is part of the pecularities of a nation which therefore is supposed to protect it and to make good use of it.

Fortunately enough, no responsible politicians will deny the importance of coordination and agreement between countries to deal effectively with environmental problems and their plan for solutions.

The dialogue between countries is not always easy, particularly when they have to deal with resources, but the discussions, the analyses and eventually the solutions could be more easily identified in multilateral fora in which the overall interest is intended to prevail on the interest of a single country.

Here it comes the important role which international organisations can play at the level of tackling environmental issues. Those organisations which have their mandate specifically in the environmental field must of course have the leading role and should steer the debate and the recommendations towards the optimal solutions for a better environment. Still other organizations, the mission of which is in other fields, could be taken as an added value in this respect. The best example on this is the ENVSEC initiative, bearing in mind that the collaboration between different players dealing with environmental issues means essentially to make available for any given initiative or programme, specific competences, skills and resources, which are essential in making more effective the engagement of the society to preserve the environment, to respect it and also to make it safe and secure for the benefit and the well being of populations.

References

Adeel Z, Safriel U et al (2005) Ecosystems and human well being desertification synthesis. millennium ecosystem assessment. World Resource Institute, Washington, DC

Blue Plan (2005) A sustainable future for the mediterranean. In: Benoit G, Comeau A (eds) The blue plan-s environment and development outlook. Earthscan, London

Brauch HG, Liotta PH et al (2003) Security and environment in the mediterranean. conceptualising security and environmental conflicts. Springer, Berlin

de Kabermatten G (2009) Grounding security, securing the ground. In: Rubio JL, Safriel U (eds) Water scarcity, land degradation and desertification in the mediterranean region. Springer Science and Business Media, The Netherland

Myers N (2004) Environmental security: what's new and different? Background paper for The Hague conference on Environment, Security and Sustainable Environment

Pedrazzini F (2008) Water scarcity, land degradation and desertification as factors for social and political instability. In: Environment, Forced Migration and Social Vulnerability, International Conference Proceedings http://www.efmsv2008.org/

Renner M (2006) Introduction to the concepts of environmental security and environmental conflict. Institute for Environmental Security, The Hague. Available at http://www.envirosecurity.org/ges/inventory/IESPP_I-C_Introduction.pdf

Kiss K (2006) Report on the OECD workshop "Subsidy reform and sustainable development." Helsinki, 20–21 June 2006

Rubio JL (2009) Desertification and water scarcity as a security challenge in the mediterranean. In: Rubio JL Safriel U (ed) Land degradation and desertification in the Mediterranean region. Springer Science and Business Media, The Netherland

Schrijver N (1989) International Organisation For Environmental Security. Secur Dialogue 20(2):115–122

Thibault H-L (2009) Facing water crisis and shortages in the mediterranean. In: Rubio JL Safriel U (ed) Water scarcity, land degradation and desertification in the mediterranean region. Springer Science & Business Media, Israel

Civil Emergency Planning. A key security task for the alliance. Available at http://www.nato.int/issues/cep/index.html

NATO Science for Peace and Security Programme. Available at: www.nato.int/science

Millennium Ecosystem Assessment (2005) Ecosystems and human well being and desertification synthesis. World Resources Institute, Washington, DC

Desertification: a security threat? Analysis of risks and challenges. A Conference on the occasion of the World Day to Combat Desertification. Berlin, June 26 (2007). Available at http://www.gtz.de/

European Commission Report on activities undertaken and support provided by the European Community to combat Desertification in countries in Asia, Latin America and Caribbean, Central and Eastern Europe in the period January 2001, December 2005, March 2007

European Commission. Kopernikus: observing our planet for a safer world. Available at: http://ec.europa.eu/enterprise/newsroom/cf/ itemlongdetail.cfm?item_id=1761&lang=it

OSCE the Environment for Europe process. Available at: http://www.environmentforeurope.org/institutions/osce.html

The United Nations. University-Institute for Environment and Human Security UNU-EHS Homepage http://www.ehs.unu.edu/

ENVSEC Environment and Security Initiative. Homepage http://www.envsec.org/

UNCCD Homepage http://www.unccd.int/

Institute for Environmental Security. Homepage http://www.envirosecurity.org

United Nations Environment Programme UNEP. Homepage http://www.unep.org/

The Environment and Security Initiative in South Eastern Europe: Transforming Risk into Cooperation

Pier Carlo Sandei

Abstract South Eastern Europe, being shattered with conflicts in the last decades, still faces numerous challenges today such as inadequate growth, declining living standards and high environmental stress. Climate change will exacerbate the situation in the future. Since environmental security is an emerging concern that cannot be tackled with traditional approaches, new ways to deal with it have to be developed. Cooperation forms the core of such means, aiming at a cross-sectoral approach and guaranteeing a wide-stakeholder participation with the view of integrating the principle of sustainable development into national policies.

The ENVSEC, a unique partnership between the UN and other agencies that is targeting certain hotspots in Europe and Central Asia can serve as a model to meet security and environmental concerns also in other areas of the world.

Keywords ENVSEC • Environment • Security • South Eastern Europe • Western Balkans • Trans-boundary environmental risks • Cross-sectoral programme • Climate change adaptation • Natural resources management • Mining • Public participation • Cooperation • OSCE • UNEP • UNECE • UNDP • REC • NATO

1 Background

It is widely recognized today that the destruction and over-exploitation of natural resources and ecosystems negatively affect the socio-economic development and can threaten societal stability. Similarly, disputes over environmental issues such as cross-border pollution or shared assets such as rivers and lakes can cause political tension and even conflict. Countries experiencing an economic transition or political stress are particularly vulnerable to environmental damage and competition over

P.C. Sandei (✉)
Regional Office for Europe, UNEP, Vienna International Centre, Vienna, Austria
e-mail: PierCarlo.SANDEI@unvienna.org

resources. The Environment and Security Initiative (ENVSEC)[1] recognizes that the best path to addressing environmental and security concerns is through international dialogue and neighbourly cooperation. It therefore assists governments in identifying common solutions and in developing joint projects for achieving them.

The ENVSEC Initiative addresses the critical links and impacts of environmental management, human security, conflict prevention and sustainable development in the areas that are prone to conflicts and growing tension. By being a cross-sectoral programme that touches upon many aspects in the environment and sustainable development field, the programme is positioned to help countries pursuing integrated strategies for environmentally sound and sustainable development and poverty reduction. Its niche primarily lies in the promotion of environmental governance, in assisting countries to build cross-sectoral capacities, put in place effective policies and institutions as well as participatory mechanisms to protect the environment, prevent conflict and reduce human poverty.

By addressing areas such as trans-boundary resource management, environmental impacts on public health, environmental information and public participation, ENVSEC aims at contributing to national development frameworks that mobilize political commitment to improve local livelihoods and to reach the Millennium Development Goals (MDGs). The interlinked cross-thematic activities of ENVSEC serve to work towards several of the MDGs. However, they are particularly relevant to the targets of Goal 7, namely to ensure environmental sustainability through the integration of the principle of sustainable development into the country policies and programmes by providing assistance and removing the main obstacles for national and local sustainable development strategies and policies. ENVSEC supports cross-border capacity building and policy development activities to help national authorities report on the implementation of the environmental multilateral conventions, such as the UN Biodiversity Convention,[2] the UN Climate Change Convention,[3] and the following UNECE Conventions: the Convention on Access to Information, Public Participation in Decision-Making and Access to Justice in Environmental Matters (Aarhus Convention), the Convention on Environmental Impact Assessment in a Trans-boundary Context (Espoo Convention), the Convention on the Trans-boundary Effects of Industrial Accidents as well as the Convention on the Trans-boundary Watercourses and International Lakes and their protocols.

All ENVSEC activities focus on or include capacity building components and promote recipient country involvement in order to reach the largest possible and most sustainable results on the ground. Stakeholders and beneficiaries are taking active part in project development and implementation, thereby learning by doing.

The inter-organizational and coordinated action of the ENVSEC partner organizations (UNEP, UNECE, UNDP, REC, OSCE and NATO), with their combination of

---

[1] More information on ENVSEC can be obtained from http://www.envsec.org [last visited 7 June 2010].

[2] Convention on Biological Diversity 1992.

[3] United Nations Framework Convention on Climate Change 1992.

skills, experiences and networks, offers a unique value proposition rarely attributed to international assistance programs. Through coordinated planning among partners and other international actors, ENVSEC has the necessary visibility to focus interventions without the danger of duplication. ENVSEC work also relies on the support of the already established field presences of OSCE, UNDP and REC to quickly and cost-effectively implement a range of instruments required for catalyzing solutions to environment and security problems. Most importantly, each ENVSEC partner brings unique alliances and contacts with various government agencies including ministries of foreign affairs, defense, economic development and environment to advance the environment and security agenda in the most comprehensive manner as well as the access to different stakeholders and their networks. According to the preliminary results of the Strategic review of the ENVSEC Initiative (conducted during the first half of 2009), ENVSEC remains unique as an innovative partnership of agencies and organizations, representing complementary specializations and expertise, and offering to respond to the complex environmental and security challenges in an integrated way.

2 The Situation in South Eastern Europe

The past decades of war, conflict and transition left the South Eastern Europe (SEE) region with a legacy of inadequate growth, declining living standards and high environmental stress. The legacy of military activities in the Balkans resulted in the degradation of ecosystems due to hazardous and toxic waste (depleted uranium, landmines, and pharmaceutical waste), the destruction of the water infrastructure, the loss of institutional and administrative capacities and an increased pressure on local ecosystems due to refugees and returning populations. As a consequence of the conflict, living standards have sharply declined, evidenced by higher poverty, inequality and unemployment and limited prospects for economic growth. Even though the security situation in SEE has been stabilized the countries with the so called "fragile democracies" are still recovering from conflicts.

As stated in the ENVSEC assessment report *"Environment and Security: Transforming Risks into Cooperation: The Case of Central Asia and South Eastern Europe" (UNEP, 2003)*,[4] there are close links between security and environment in the SEE region. Specific security risk are driven by controversial environmental issues such as the management of trans-boundary risks of hazardous activities, the management of shared natural resources and the arising of novel, unexpected trans-boundary risks induced by climate change.

The SEE region is mainly affected by heavy industrial pollution in urban-industrial areas, industrial pollution from the mining sector, intensive agriculture with yet uncalculated health impacts, lack of water technology, infrastructure as well as industrial pollution from the mining sector. In addition, use and management of

---

[4] Available under: http://www.envsec.org/docs/envsec_eastern_europe.pdf [last visited 07 06 2010].

shared natural resources such as trans-boundary lakes and rivers, mountain forest complexes as well as biodiversity (e.g. in the Dinaric Arc and Balkan mountains) pose both a challenge and an opportunity for cooperation.

Multilateral environmental agreements provide governments with the necessary guidance for addressing most of these challenges. However, their enforcement in many of the countries of the region is still far from being satisfactory. Their entry into force and implementation demand high political willingness as well as innovative and swift action beyond conventional measures, both of which require informed and active civil society engagement as the primary driving force. Nevertheless, the countries involved in the proposed project did not have environmental issues among the top priorities of their past political agendas as all of them were afflicted by the Yugoslav Wars in the ending twentieth century. Moreover, the violent conflicts worsened the barriers of economic growth posed by the isolation from global economy and the impacts from the transition of a planned socialist economy to a free market economy. The low economic growth resulted in cheap operating principles, which are very hazardous to the environment as for example unsustainable mining techniques that induced serious pollution throughout the region, as well as unsustainable agriculture, hunting and forestry.

In addition, in its latest assessment report of 2007,[5] the Intergovernmental Panel on Climate Change (IPCC) projects a worldwide severely changing climate with a simultaneously increasing summer dryness and an increased chance of intense precipitation for northern, middle and high latitudes of the SEE region. This means that, with the utmost probability, future summer precipitation in the SEE will concentrate to fewer and more intense events interrupted by longer dry periods, thus enhancing the risk of intense soil erosion and severe forest fires. The changing climate will also enhance the pressure on biodiversity loss and aggravate the effects deriving from mining pollution. The heavy metal concentration of polluted waters will increase in drier periods, which will be extremely harmful for agriculture. Furthermore, coupled with an increased frequency and magnitude of natural hazards (e.g. floods, droughts, soil erosion, forest fires), climate change is likely to weaken the fragile economies of the affected countries. The extreme weather phenomena and natural hazards might also evoke new tensions among the states, for example with regard to water scarcity or the responsibilities for remediation of the affected people.

3 Priorities and Activities

The focus of the ENVSEC work in the SEE lies within the Initiative's three main pillars[6]:

- promoting in-depth vulnerability assessment, early warning and monitoring of environment and security risks;

---

[5] IPCC Fourth Assessment Report (AR4), available under: http://www.ipcc.ch/publications_and_data/publications_ipcc_fourth_assessment_report_synthesis_report.htm [last visited 07 06 2010].
[6] See also http://www.envsec.org/see/index.php [last visited: 07 06 2010].

- improving awareness on the interrelation between environment and security, strengthening environmental policies, and improving the capacities and the roles of environmental institutions;
- providing technical expertise and mobilizing financial support for clean-up and remediation

The priority fields of action for SEE that were developed jointly by ENVSEC partners and national experts in close co-operation with the national representatives of the relevant governments countries during regional consultations comprise:

- management and reduction of trans-boundary risks from hazardous activities;
- management of shared natural resources;
- strengthening regional cooperation on environmental governance through participatory and informed decision making and implementation processes;
- adaptation to the impacts of climate change for reducing security risks in SEE.

These above mentioned priorities are interrelated and in particular the cross-cutting priorities of climate change and awareness raising/public participation greatly contribute to the implementation of the other two priorities on hazardous activities and management of shared natural resources.

The results of the past and ongoing ENVSEC projects carried out in SEE[7] include, among others, the identification of hotspots that pose substantial trans-boundary risk to the environment, the increase of local capacity for early warning, conflict resolution and emergency response, an improved information base on the state of biodiversity as well as improved overall cooperative mechanisms.

4 ENVSEC Activities in See in the Period 2009–2012

4.1 Trans-boundary Mining Hotspots

The mining sector is an important contributor to local and national economies in SEE. However, in various parts of the region, mining is often characterized by inappropriate planning, as well as operational and post-operational practices taking place within inadequate regulatory frameworks. Poor or negligible implementation of mine rehabilitation and closure activities have been performed so far. In SEE this

---

[7] During the period of 2006–2009, the implementation of activities related to these priorities were mainly supported by the Austrian Development Agency (ADA) through the project Environment and Security in South Eastern Europe: Improving regional cooperation for risk management from pollution hotspots as well as the transboundary management of shared natural resources and the Canadian International Development Agency (CIDA). The activities, implemented by UNEP's ROE through its Vienna Office provided in-depth regional assessments for both mining hotspots and transboundary biodiversity, as well as site-specific feasibility studies and projects to be built upon and followed up in the further work of the initiative.

has resulted in, and continues to cause, significant adverse environmental, health and safety impacts and related liabilities. Active and abandoned mining sites belong to the most widespread environmental concerns across the Western Balkans.

Chemical sites in the region should also be given a greater attention, so that the SEE countries could strengthen the prevention, preparedness and response to industrial accidents owing to better risk management, relevant safety measures and the establishment of contingency planning.

Trans-boundary environmental problems, such as water, air and soil pollution from toxic/acidic effluents, airborne toxics and so on arise from substandard operations and improper mine closure. Across the region, poor conditions of mining facilities, such as tailings dams, waste heaps and so on, have caused severe pollution mostly by releasing heavy metal contaminating waters. The chronic pollution does not only affect plant and animal life, but also human health and the economic development of the region, as the strong negative media presence linked to such spill accidents regularly repels tourists and potential purchasers of local products. Trans-boundary effects of such pollution can trigger substantial tensions of the still feeble political cooperation in the region and might affect the security of the Balkan states.

Watercourses are the main vector for trans-boundary pollution, whether it is ongoing and chronic, or infrequent and accidental. Discharges with a low pH value and rich in heavy metals affect downstream ecosystems and make water unsuitable for irrigation and other purposes. Many waterways cross the borders and as the countries are relatively small, many sites are located close to a neighbouring state.

ENVSEC intervention focuses on the prevention and mitigation of trans-boundary environmental risks arising from hazardous pollution hotspots in particular from abandoned mines, dams and chemical sites as well as capacity building to support countries to ratify and implement the Industrial Accidents Convention, the Espoo Convention and its SEA Protocol and the Water Convention.

This leads to a reduced local environmental and human health risks and minimized tensions among SEE countries through the reduction of risk of accidental trans-boundary pollution arising from hazardous mining sites and to an improved trans-boundary and regional cooperation on identification, management and reduction of trans-boundary risks from hazardous activities.

4.2 Management of Shared Natural Resources

Europe is characterized by many borders that cut across ecosystems and areas of high natural values, often dividing the continent along natural barriers like mountain ranges or rivers. Border areas are often the most favoured regions in terms of biodiversity, partly as a result of their peripheral location or political factors banning in the past the development of areas adjacent to political borders.

However, natural areas shared by neighbouring countries are not only a common treasure, but also a common responsibility. Therefore, achieving the ecological coherence of Europe, protecting and managing its natural resources in a sustainable way, as well as preventing or mitigating environmental threats cannot be achieved by one country alone and require inter-regional and trans-boundary cooperation.

Trans-boundary co-operation on shared natural resources represents an important tool to mitigate the adverse environmental impacts on the economy and health of affected communities and to explicitly create trust and confidence among nations, which previously experienced political tensions and even violent conflict.

To encourage, enhance and support trans-boundary and regional cooperation of governments and local stakeholders on management of shared natural resources of the countries, ENVSEC will particularly focus on selected trans-boundary mountain protected areas with an ecosystem services based approach, trans-boundary rivers and illegal logging.

This will lead to a better understanding of the interrelated services provided by ecosystems of selected regions, looking in particular at (high) trans-boundary eco-regions and mountain environments (e.g. Dinaric Arc, Balkans), to an improved cross-border and regional dialogue for cooperation on the sustainable development of the region and promotion of formal collaborative agreements as well as to an enhanced management of shared trans-boundary mountain ecosystems and cooperation for the establishment of trans-boundary mountain protected areas

4.3 Public Participation

Taking into account the effort to address environment and security challenges, particularly in the trans-boundary context, the ENVSEC partners' work has been instrumental in promoting Multilateral Environmental Agreements (MEAs) in the SEE region and in establishing networking among a variety of stakeholders, including legal professionals.

These awareness-raising activities have further fed into some pilot level initiatives such as promoting public participation in the EIA process and in the establishment of an Aarhus Centre in Albania. These initiatives demonstrated the significant need for organized support towards building the capacities of national stakeholders in participatory environmental governance, with particular focus on regional/local administrations and civil society organizations.

This priority area will mainly focus on strengthening the capacities of civil society and raising awareness on its role in environmental protection, conflict prevention and resolution, through trainings, stakeholder meetings, public hearings and other tools that facilitate its active participation in environmental decision-making processes. It will also aim at increasing the access to environmental information by the civil society and public at large in order to improve dialogue and public awareness.

Priority will be given to the implementation of the Aarhus and Espoo Convention and their related protocols, as well as to activities closely related to the management of shared natural resources, the management of risks and pollution and climate change issues.

4.4 Climate Change Adaptation

As stated in the recently published report of the United Nations Secretary-General on "Climate change and its possible security implications" (September 2009),[8] climate change can act as a 'threat multiplier', exacerbating threats caused by persistent poverty, weak institutions for resources management and conflict resolution, a history of mistrust between communities and nations, and inadequate access to vital natural resources such as water and arable land. The inability to cope with these situations may result in potentially violent conflicts.

Natural hazards, loss of biodiversity and according ecosystem services, as well as degradation of landscape and their implications for the socio-economic sector might cause serious setbacks for the SEE region's economic growth and development. The adverse effects of climate change thus pose a serious threat to the still feeble trans-boundary collaboration and mutual confidence, as well as to the sustainable development of the region.

To tackle the arising challenge that climate change bears, the lessons learned from other ongoing projects and activities may ensure the further success of ENVSEC activities in this sector.

In this sense, it should be noted that five of the target countries have already agreed upon the "South East European Climate Change Framework Action Plan for Adaptation" (SEECCFAPA),[9] hence recognizing the urgency to take action in the face of climate change and have shown their determination to further cooperate.

5 Conclusion

The linkages between environment and national and international security are widely recognized in the international context although they are often hidden and denied by countries and relevant national stakeholders. In this situation the very pragmatic and action-oriented approach of the ENVSEC initiative has proven to be effective and well accepted by the countries in SEE. The motto *"Transforming risk into Cooperation"* is an effective way to describe the way ENVSEC works in this field.

The ENVSEC partnership with its different constituencies has demonstrated its value in mobilizing the different stakeholders and in building a strong national support to the initiative, which resulted in a wide public participation and bottom-up approach in the definition of future activities.

---

[8] Available under http://www.un.org/ga/search/view_doc.asp?symbol=A/64/350 [last visited 07 06 2010]. The report of the UN Secretary General was delivered in response to General Assembly Res. A/63/L.8, adopted on 27 October 2008, linking the issues of climate change and security implications.

[9] Available under: http://www.rcc.int/download.php?tip=docs&doc=Executive+Summary.pdf&doc_url=880018115ffd8639b185f2a49ba452ab [last visited: 07 06 2010].

In conclusion, the ENVSEC initiative remains a unique partnership between UN and non-UN organization which can represent a model for other regions of the world facing similar problems.

References

Agency of Environment and Forests of Albania (2008) State of the environment report 2005–2006 and 2007
Albania National Biodiversity Strategy and Action Plan (2000) Tiranë
Albania National Environment Strategy (2007) Tiranë
Biodiversity Enabling Activities Book – Albania (2007) Article II and Article III
Brininstool M (2007) The Mineral Industry of Serbia, USGS
Council of Europe (2007) Emerald project for the setting-up of the network of areas of special conservation interest in Albania (2nd phase 2005–2006)
Demi G (2007) Waste assessment of Copper mines and plants in Albania and their impact in surrounding areas. Mining and Processing Technology Institute, Tirana
Diku A (2008) Albania country report for UNEP Vienna on the feasibility of establishing a transboundary protected area. ILIRIA, Tiranë
European Commission, Directorate-General for Economic and Financial Affairs (2009) The Western Balkans in transition, Occasional papers 46
Foldessy J (2004) Towards sustainable ore mining in central and Eastern Europe – The contribution of an information network about mining environmental technologies
International Finance Cooperation (IFC) (2007) Environmental, health and safety guidelines for mining
Krasniqi E (2009) Kosovo report for UNEP Vienna on the feasibility of establishing a transboundary protected area. Department of biology, Faculty of natural sciences and mathematics, University of Prishtinë, Prishtinë
Ministry of Environment of the Republic of Albania (2002) The first national communication of the Republic of Albania to the United Nations. Framework Convention on Climate Change (UNFCCC), Tirana
Ministry of Environment, Forests and Water Administration of Albania (2007) Protected areas of Albania book. Tirane Agriculturale University, Tiranë
Ministry of Environment and Spatial Planning of Kosovo – KEPA (2008) State of environment report 2006 – 2007. KEPA, Prishtinë
Ministry of Environment and Spatial Planning of Kosovo – KEPA (2008) State of nature report 2006 – 2007. KEPA, Prishtinë
Ministry of Environment and Spatial Planning of Kosovo – Kosovo Institute on Nature Protection (2005) Vlerat e Trshëgimisë natyrore të Kosovës (Kosovo heritage nature values book). Prishtinë
Mudd G (2008) Sustainable mining – An oxymoron?
Niewiadomski Z (Ed) (2006) Enhancing trans-boundary biodiversity management in South Eastern Europe. Report prepared under the environment and security initiative, UNEP Vienna
Pajevic A (2007) Report of the II phase of the project "Establishing Emerald network in Montenegro". Ministry of tourism and environment, Podgorica
Peck P (2004) Reducing environment & security risks from mining in South Eastern Europe
Peck P (2005) Mining for closure - policies, practices and guidelines for sustainable mining and closure of mines. UNEP
ENVSEC (2004) Desk-assessment study
Shala D (2002) Eco-guide: nature of Rugova – a capsized wealth. Aquila Environmental Protection Association, Peja
Steblez W (2006) The mineral industries of the Southern Balkans. USGS

UNECE (2003) Environmental performance review of Kosovo

ENVSEC (2007a) UNEP/GRID Arendal; Balkan vital graphics –Environment without borders

ENVSEC (2007b) UNEP/GRID Arendal; innovative techniques and technologies for contaminated mine waters assessment, management and remediation

UNEP Vienna ISCC (2009) Feasibility study on establishing a trans-boundary protected area Durmitor - Tara Canyon – Sutjeska. Vienna

UNEP Vienna ISCC (2009) Feasibility study on establishing a trans-boundary protected area Sharr/Šar Planina – Korab – Dešat/Deshat. Vienna

USAID / ARD-BIOFOR IQC Consortium (2003) Kosovo biodiversity assessment. Prishtinë

Various Authors (2004) Environment and security consultations in South Eastern Europe, Final report, Skopje, FYR of Macedonia, 23–24 Sept 2004

Warhurst A (1998) Mining and the environment: case studies from the Americas, International Development Research Centre (Canada)

Western Balkans Environmental Program (WBEP) (2009) Strengthening capacities in the Western Balkans countries to mitigate environmental problems through remediation of high priority hotspots. 30 month, 15 million USD Netherlands Government/UNDP program – www.westernbalkansenvironment.net

World Health Organization (WHO) (2006) Guidelines for drinking-water quality, 3rd edn, incorporating first and second addenda

The Role of UNESCO Designated Sites in Fostering International Cooperation and Environmental Security in SEE [*]

Giorgio Andrian and Karl-Heinz Gaudry

Abstract In the rapidly changing scenarios of the South-Eastern Europe, control on territories has always been playing a relevant geo-political role. In particular, the recent and relatively rapid modification of the regional geography – within which the break apart of the *Socialist Federal Republic of Yugoslavia* was the most significant event – resulted in the creation of a high number of 'new' borders, with the direct consequence of multiplying the trans-boundary initiatives. In many cases, unsolved border issues are still pending, keeping a high conflict potential. Under these conditions, the role of the international initiatives – and in particular, the activities carried out within various multilateral agreements – prove to be powerful tools to reconcile part of these tensions. A specific example is represented by the UNESCO designated sites, namely, the *World Heritage Sites* and the *Biosphere Reserves*. In both cases, their international visibility and respective governance mechanisms guarantee to foster the multilateral cooperation. The various existing sites in SEE and, most of all, the ones to be designated next, represent very important opportunities to 'ground' the trans-boundary cooperation and to potentially catalyze the broad range of current 'green diplomacy' activities in the region, fostering the overall environment security.

Keywords Designated sites • UNESCO • Trans-boundary cooperation • World heritage • Biosphere reserves • Environmental diplomacy • Geo-politics

---

[*] The opinions expressed in this paper do not reflect the official UNESCO position, but exclusively the authors' point of view.

G. Andrian (✉)
UNESCO Venice Office, Regional Bureau for Science and Culture in Europe (BRESCE)
Venice, Italy
e-mail: g.andrian@unesco.org

K.-H. Gaudry
Albert-Ludwigs-Universität Freiburg, Institute for Landscape Management,
Tennenbacher Str. 4, 79106 Freiburg, Germany
e-mail: karl-heinz.gaudry@landespflege.uni-freiburg.de; khgaudry@gmail.com

1 Introduction

"*Your environment, your move*", titled the workshop recently organized by UNDP[1] Montenegro in Budva,[2] within the '*Western Balkans Environmental Hotspot Programme*'. The web-published report stresses that "more than a decade after armed conflicts, six countries/territories in the Western Balkans are now working together to tackle regional environmental hotspots".[3]

The tumultuous recent history of the Balkan Peninsula is calling for more international cooperation to foster regional stability; in this context, environmental issues are playing a growing role as issues of high geopolitical relevance.

The role that the international community is asked to play is of crucial importance in creating the necessary neutrality and multilateralism that are essential preconditions for any successful negotiation. In this particular context, the United Nations[4] – and its specialized agencies' system – prove to be effective in promoting the adoption of integrated approaches at various scales (from the global to the local), within a broad spectrum of 'environmental diplomacy' activities.

Ultimately, the need to localize on the map the sites of common interest where to concentrate the local and the sovra-local efforts, offer a unique chance to 'ground' the policy efforts and to 'test' the validity of integrated approaches.

Within this scenario, the UNESCO designated sites – namely, the *World Heritage Sites* and the *Biosphere Reserves* – represent examples of outstanding values (both natural and cultural) with international recognition,[5] that may well serve the purpose of fostering the trans-national cooperation.

This paper – by providing the *state-of-the-art* in the specific domain of the UNESCO related activities in SEE – aims at illustrating the potential of the internationally designated sites in fostering environmental and security in the region, and, ultimately, contributing to filling a still perceived research gap.[6]

2 South-Eastern Europe: A Constantly Changing Geo-Political Scenario

Recalling the metaphor used by Castoriadis[7] to signify the thinking process, approaching the SEE region is like 'entering the labyrinth'. Starting from the acronym – which stands for *South-Eastern Europe* – that has been largely adopted by the international community, as the most acceptable and 'politically correct' identification of that region; in fact, the

---

[1] The United Nations Development Programme (www.undp.org).
[2] Budva, Montenegro, November 19, 2009.
[3] http://www.undp.org.mk/default.aspx.
[4] http://www.un.org/.
[5] 'UNESCO Biosphere Reserves: model regions with a global reputation', titles the issue 2/2007 of 'UNESCO today', the Journal of the German Commission for UNESCO.
[6] Jones 1994, p 316.
[7] Cornelius Castoriadis, *Crossroad in the Labyrith*, Brighton: Harvester Press, 1984, pp-ix–x.

Fig. 1 *South East Europe* according to the SEE transnational cooperation programme (SEE-TCP 2010)

term 'Balkan'[8] – and its related adjectives – still incorporates a very negative connotation, and has, consequently, been rejected in the official international relations terminology.

In fact, there's no consensus on the 'borders' of the SEE region: different maps (see Fig. 1 and Fig. 2) are used by the various international organizations, where, by the same acronym, different countries are included. Similarly, the use of the term 'Western Balkans' has generated controversial reactions among the various included/excluded countries.

Regardless the recent 'tentative geographies' of the region, the fact that the Balkan Peninsula has seen several important 'internal' borders changes during the last centuries is a matter of fact. The Fig. 3 well illustrates these major changes, showing how the relatively long periods of multinational empires (namely, the Ottoman and the Austro-Hungarian) dominations has been brought to an end while approaching the beginning of the twentieth century. After the so called 'First' and 'Second' Balkan wars'[9] – in 1912 – 3 – the Ottoman Empire domination had vanished and being replaced by a number of independent States.

---

[8] For a complete overview on the origin and evolution of the term 'Balkan', see Maria Todorova, *'Imaging the Balkans'*.

[9] For a detailed description of the recent Balkans history, see Hösch 2004, *Geschichte des Balkans*.

Fig. 2 The presence of the UN- Office for the coordination of humanitarian affairs in "South Eastern Europe" (ReliefWeb 2000)

After the Treaty of Lausanne (1923), the Yugoslavia appeared on the map; after the Second World War (1943) the *Socialistic Federal Republic of Yugoslavia* was formed and for the following 45 years it played a dominant role in the region and among the non aligned countries. Its dissolution–officially started by the proclamation of the independence of Slovenia and Croatia–was followed by a long and cruel series of armed conflicts that ended only after 1999.

The current picture sees six recently established republics instead; in addition, the self-proclaimed independent Kosovo[10] is bringing one more piece in the Balkan 'puzzle'. These relatively rapid state-building processes brought the extension of the national border to grow tremendously, with all the consequences related to the cross-border issues. Additionally, some borders issues are still to be solved; the example of the disputes between Croatia and Serbia along the Danube[11] is meaningful of the potential threats to foster cross-border cooperation in post-conflict situation.

Parallel to the dissolution of the former Yugoslavia, the SEE countries started individual processes of European Union approximation; Romania and Bulgaria

---

[10] The proclamation of independence of Kosovo (February 2008) has not been recognized by the United Nations and many of its Members, creating a pending status definition in relation with the Republic of Serbia.

[11] This issue emerged on the occasion of the organization of the UNESCO workshop on the '*Danube-Drava-Mura Transboundary Biosphere Reserve: scenario of integrated river management*', held in Belgrade and Novi Sad, July 23–25, 2009; Croatian authorities claimed that the border between the two protected areas of 'Copački Rit' (in Croatia) and 'Gornje Podunavlje' (in Serbia) has not been clarified yet, impeding any possible transboundary cooperation in that part of the country.

The Role of UNESCO Designated Sites in Fostering International Cooperation

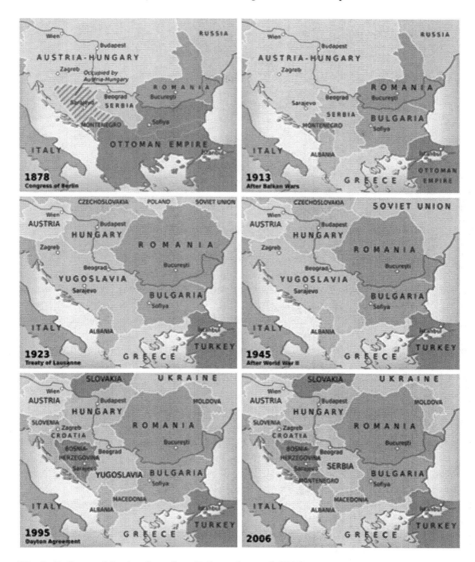

Fig. 3 Balkan multinational empires (Rekacewicz et al. 2007)

were given the full membership in 2007, while the other republics are facing different 'speeds' in the process (Fig. 4).

Despite the fact that – as many scholars and international observers agree on – "Europe as an emerging global player critically viewed against the backdrop of challenges posed by the delayed transition in the Western Balkans" Swoboda and Solioz (2007), the European integration represents the far most important priority in all the non-EU member current governments strategies.

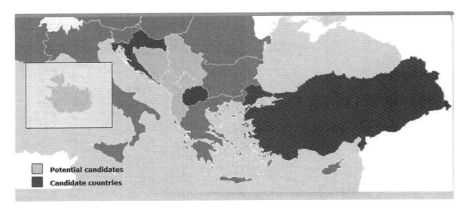

Fig. 4 Countries as *"potential EU candidates"* and *"EU candidate"* (EC 2010)

Beside the one-to-one negotiating rounds opened by the EU with each of the SEE country, more integrated and trans-boundary initiatives are promoted. Under the slogan *'jointly for our common future'* the *South East Transnational Cooperation Programme* has been launched (see the eligible territories in Fig. 1) and co-funded by the European Union: it intends to be a "unique instrument which, in the framework of the *Regional Policy's Territorial Cooperation Objective*, aims at improving integration and competitiveness in an area which is as complex as it is diverse".[12] In line with the Lisbon and Gothenburg priorities, the Program also intends to contribute to the integration process of the non-EU member states, by supporting projects developed within four Priority Axes: *Innovation, Environment, Accessibility,* and *Sustainable Growth Areas.*

It is interesting to note how *environment* is considered one of the strategic axes to foster the international cooperation. Driven by the basic assumption that *'Nature unites what borders divide'*,[13] many opportunities are offered in the region by the existence of large ecosystems that are across the recently established national borders: long rivers (e.g. Danube, Sava), large lakes (e.g. Skadar/Shkodra, Prespa), vast mountain chains (e.g. the Dinaric Arc) represent reasons to adopt the eco-regional approach[14] in the attempt to overcome the constraints given by the administrative borders.

3 UNESCO Designated Sites: Brief Historical Development and Governance Mechanisms

Within the vast range of *Multilateral Environmental Agreements* (MEAs) that are currently in force, only very few include the identification of specific territories; namely, the *Convention on Wetlands of International Importance* (1971, better

---

[12] http://www.southeast-europe.net/hu.

[13] This is a famous slogan created by IUCN to foster transboundary cooperation along the *European Green Belt* (for more details, see http://europeangreenbelt.org/).

[14] An interesting example is given by the *Dinaric Arc Ecoregion*, as defined by WWF (for details, see: http://www.panda.org/).

known as *Ramsar Convention*), the *Convention Concerning the Protection of the World Cultural and Natural Heritage* (1972, better known as *World Heritage Convention*), the *Man and Biosphere Programme* (MaB) and its *World Network of Biosphere Reserves*. The last two are coordinated by UNESCO, which hosts their respective Secretariats.

All of them represent long lasting and well established intergovernmental initiatives that, from the original mission to identify and protect locations of global importance, have become very useful tools to foster the international cooperation. The case of the UNESCO designated sites – the *World Heritage*, in particular – signifies the importance of the international visibility as a key driver to foster the *Member States* in taking seriously the protection and proper management of the listed territories.

3.1 The Convention Concerning the Protection of the World Cultural and Natural Heritage (the World Heritage Convention)

Regarded by many as the 'Nobel Prize for Nature', the *World Heritage Convention* provides identification and protection to sites of well recognized "outstanding universal value".

It is based on the core idea of identifying those locations around the world which are of outstanding universal *natural* and *cultural* values[15], to be properly protected and managed. Entered into force 1972, the Convention[16] recognizes in its art. 4, that each *Signatory Party* has the duty of ensuring the *identification, protection, conservation, preservation* and *transmission to future generations* of the cultural and natural heritage – as defined in art. 1 and 2. The original idea of combining the conservation of cultural sites with those of nature comes originally from the USA. A *White House Conference* in Washington, D.C., in 1965, stimulated the international cooperation to protect the world's superb natural and scenic areas and historic sites for the present and the future of the entire world citizenry. This proposal was presented to the 1972 *United Nations Conference on Human Environment* in Stockholm; once a large agreement was reached, the Convention concerning the *Protection of World Cultural and Natural Heritage* was presented and ultimately adopted by the General Conference of UNESCO on November 16, 1972. The creation of a specific *List* (the *World Heritage List*) has strongly characterized the *Convention* related activities, being the inscription of new sites the key mechanism in place to implement its principles. The criteria in accordance to which a site can be proposed for the inscription are well listed and described in the *Operational*

---

[15] The concept of *outstanding universal value* is defined in Par. 49 of the *Operational Guidelines for the implementation of the World Heritage Convention* as following: "Outstanding universal value means cultural and/or natural significance which is so exceptional as to transcend national boundaries and to be of common importance for present and future generations of all humanity. As such, the permanent protection of this heritage is of the highest importance to the international community as a whole".

[16] For further details, see the dedicated web site: http://whc.unesco.org/en/convention.

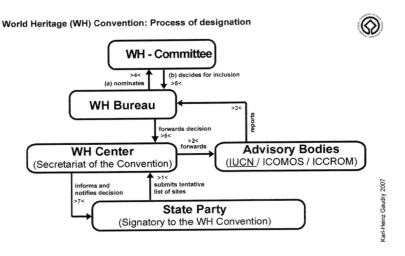

Fig. 5 The convention's designation governance mechanism in (Gaudry 2007)

Guidelines for the implementation of the World Heritage Convention, lastly updated in 2008.[17]

Since its entering into force, a constantly growing trend in site designation number has been recorded, bringing the current figure to a total of 890 *properties* – 689 *cultural*, 176 *natural* and 25 *mixed*.[18] The worldwide recognition of the Convention's importance is reflected by both the very high number of its *Signatory Parties* – as of April 2009, 186 UNESCO *Member States* ratified it – and by its tremendous implementation rate – *World Heritage sites* are found in 148 countries.

The designation process is bottom-up (see Fig. 5), starting when a *Signatory Party* decides to propose a site on its territory to be listed (step 1). To encourage countries to activate a proper selection procedure, *Tentative Lists* were created by every country to include the sites intended to be listed: a minimum 1 year period is asked for tentative candidatures to wait before being officially proposed, in order to be fully analyzed and compared with other similar sites of outstanding values that are present around the world. Every year, a deadline is established by the *World Heritage Centre* (acting as *Secretariat* of the Convention) to accept official dossiers from the States. External *Advisory Bodies*[19] are asked to revise the official candidatures (step 2), before being forwarded to the *World Heritage Bureau* (step 3). The ultimate decision making body is the *World Heritage Committee* (step 4), an organ composed by 21 representatives of the *State Parties*, elected by the UNESCO General Assembly. On the occasion of its annual meeting, the *Committee* analyses

---

[17] http://whc.unesco.org/en/criteria.

[18] See http://whc.unesco.org/en/list.

[19] The Advisory bodies are IUCN, for the natural sites and ICOMOS and ICCROM for the cultural sites.

all the accepted candidatures and take all the relevant decisions, including recommendations on the already inscribed properties, on the basis of all the officially received supporting documents. *State Parties* are required to submit relevant information on the administrative and legislative provisions taken for the application of the *Convention*; additionally, every 6 years a detailed report on the *state of conservation* of the sites is requested. Ultimately, the *Committee* outputs are then transmitted back to the *Signatory Parties* via *Bureau* and *Centre* (steps 5 and 6).

In addition to the consolidation of the Convention's role in providing the necessary support for a proper safeguarding of the inscribed sites – by promoting immediate actions and international campaigns[20] – the *Committee*'s decisions and recommendations are more and more encouraging the parties to use this tool as an instrument to foster the international cooperation.[21] In fact, the inscription of *trans-boundary sites* and *serial sites* is seen as a concrete good example in this challenging direction.

3.2 Man and Biosphere Programme (UNESCO's MaB Programme)

Launched in 1971 as an intergovernmental programme by UNESCO, the *Man and Biosphere Programme* (MaB) dates back its very origins in 1968, when the "Intergovernmental Conference of Experts on the Scientific Basis for the Rational Use and Conservation of the Resources of the Biosphere" (shortened to "The Biosphere Conference") was organized, in collaboration with FAO and IUCN.

Initially, the concept of *Biosphere Reserve* (BR) appeared in 1971, when the idea of establishing a *World Network of Biosphere Reserves* as places where combining conservation and research, was formalized. In accordance with the MaB *Statutory Framework*, BRs are defined as "areas of terrestrial and coastal ecosystems or a combination thereof, which are internationally recognized within the framework of UNESCO's programme on Man and Biosphere (MaB)".[22]

Differently from the *Word Heritage*, no Convention has been created within the MaB: in fact, the *Statutory Framework* – adopted in combination with the *Seville Strategy*, in 1995 – serves the purpose to set the guiding principles and the ultimate purposes of BRs. A *World Network of Biosphere Reserves* has been designed in order to enhance the effectiveness of individual BRs and strengthen cooperation, understanding and communication at the regional and international levels.[23]

---

[20] See http://whc.unesco.org/en/107.

[21] As recently reiterated by Mr. F.Bandarin, *World Heritage Centre* Director, om the occasion of the meeting of the Italian *World Heritage cities* (Florence, 2008).

[22] See: http://portal.unesco.org/science/en/ev.php-URL_ID=7661&URL_DO=DO_TOPIC&URL_SECTION=201.html.

[23] The operation of the Network is supported by regional and/or thematic networks such as, AfriMAB, ArabMAB, CYTED, EABRN, EuroMAB, IberoMAB, PacMAB, REDBIOS, eaBRnet and SACAM, working mostly with the UNESCO Regional Offices.

During the first years of the Program's existence, proposed BRs have been largely corresponding to pre-existing protected areas, being the initial interpretation of the MaB primarily focused on *conservation* (mostly of genetic resources and biological diversity). In 1995, on the occasion of the *Seville Conference*, a complete revision of the MaB and its functioning mechanisms was carried out: the adoption of the *Seville Strategy* and the *Statutory Framework of the World Network* marked the separation line between the so-called 'first' and 'second' generations.

As a consequence, more attention has been given to the other two major BRs functions: the *development* and *logistic support*. 'Biosphere Reserves are not just protected areas' reads the slogan of the newly proposed sites, signifying that the *core zones* were considered less extended when compared with the *buffer* and *transitions* ones.[24] Fostering economic and human development – which has to be socio-culturally and ecologically sustainable – become equally important as identifying areas of exclusive nature conservation. The 'logistics' function refers to the support for demonstrating-projects, environmental education and training, research and monitoring related to issues of local, regional, national and global importance. Ultimately – and more recently – BRs have been designated as 'learning sites', to signify the importance of the original scientific approach that is always expected to drive the adoption of innovative management procedures within those territories.

Similar to the process leading to the designation of the *World Heritage* sites, *Member States* (see Fig. 6) propose suitable territories to become BRs (steps 1 and 2): in this case, no limit in number and in time is given to the proposals to be received at the MaB Secretariat (step 4).

Certainly less known than the *World Heritage Convention*, the *MaB Programme* still represents a significant platform of designated territories related activities.

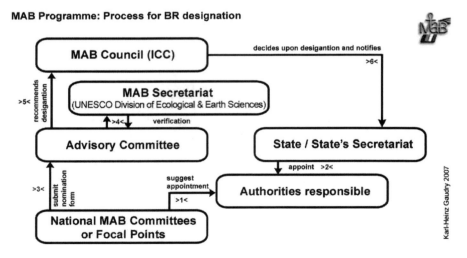

Fig. 6 The MaB biosphere reserve's governance mechanism in (Gaudry 2007)

---

[24] *Core*, *buffer* and *transition*, are the three major component of the BR zoning, according to a decreasing degree on protection.

The number of the sites has been constantly growing, reaching the current (2009) figure of 553 sites in 107 countries.[25]

It is important to note that an increasingly relevant element of the work of UNESCO – and its *Division of Ecological and Earth Sciences* that hosts the *MaB Secretariat* – is that of coordinating its activities with those of other major MEAs Secretariats, in order to ensure greater complementarities and synergies. More specifically, important working relationships have been developed with the other four biodiversity-related conventions.[26]

4 Environmental Diplomacy in Action: Current Challenges in Trans-boundary See Scenarios

Within the current environmental initiatives' map in SEE, more and more UNESCO designated sites are used as preferred locations for testing innovative policies (e.g. those related to the climate change) and implement integrated management approaches. Not only they represent 'model regions of international reputation' but also sites at the disposal of the *Member States* to fulfill their obligations derived by the various adopted MEAs.

4.1 The World Heritage and the Biosphere Reserves Scenarios

In addition to the already established ones, more *World Heritage* sites and BRs are included in the SEE countries *tentative lists*: interestingly, many of them are trans-boundary.

Three examples are of particular interest in this respect; namely, the *Prespa Lake*,[27] the *Tara River Basin*[28] and the *Danube-Drava-Mura* river corridor.[29]

"Three countries, two lakes, one future" titled the UNDP article[30] on the next initiatives devoted to the Prespa Lake: "In Support of the forthcoming Copenhagen Summit, the Prime Ministers of Albania, the former Yugoslav Republic of Macedonia and Greece, met in Prespa to sign a joint statement expressing their readiness for further promotion of environmental issues in the region. They announced that the signing

---

[25] See: www.unesco.org/mab.

[26] Namely, the Convention on Biological Diversity (CBD 1992); the Convention on International Trade in Endangered Species of Wild Fauna and Flora (CITES 1973); the Convention on the Conservation of Migratory Species of Wild Animals (CMS or Bonn Convention 1979); the Convention on Wetlands of International Importance (Ramsar 1971).

[27] The *Prespa Lake* is shared by Albania, Former Yugoslav Republic of Macedonia and Greece.

[28] The *Tara River Basin* is the official name of a *Biosphere Reserves* that encompasses large part of the Northern Montenegro territories, at the border with Bosnia and Herzegovina.

[29] The *Danube-Drava-Mura river corridor* is the tentative name given to a transnational initiative – involving five countries (Austria, Slovenia, Croatia, Hungary and Serbia) – leading to the establishment of a *Trans-boundary Biosphere Reserve*.

[30] http://europeandcis.undp.org.

of a *Tripartite Agreement for Sustainable Development of the Prespa Basin* will take place on February 2, 2010 – the 10th Anniversary of the *Trans-boundary Prespa Park"*. In this case, the establishment of the *Prespa Park Coordinating Council (PPCC)* – a well functioning tri-lateral mechanism in place since 2000 – brought its members and the other stakeholders to discuss on the possibility to use the MaB *Trans-boundary Biosphere Reserve* as the most suitable framework to properly tackle with the already existing activities into a well recognized international designation.

On the occasion of the 2nd sub-regional meeting *'Trans-boundary Cooperation of Mountain Protected Areas in SEE: Towards the Dinaric Arc and Balkan Network of Mountain Protected Areas'*, organized by the UNEP Vienna Office in the framework of the project *"Environment and Security in South Eastern Europe: Improving regional cooperation for risk management from pollution hotspots as well as the trans-boundary management of shared natural resources"*, a specific workshop was dedicated to the *'priorities for common actions in trans-boundary 'areas in focus'*. Once more, the idea of identifying some key sites of common interest proves to be one of the most effective approaches in fostering the international cooperation.

On that particular occasion, the need of improving the trans-boundary cooperation between Montenegro and Bosnia-Herzegovina 'materialized' into a concrete proposal: that of connecting the existing National Parks on the two countries – namely, 'Durmitor' in Montenegro and 'Sutjeska' in BiH. The existence of the *Tara River Basin Biosphere Reserve* in the first country, offered the idea to the national and international stakeholders to propose an extension over the border, creating a *Trans-boundary Biosphere Reserve* that would encompass both the protected areas and the territories in between.

"Danube-Drava-Mura Trans-boundary Biosphere Reserve: scenarios for integrated river management" was the title chosen by the UNESCO Venice Office to convene a workshop[31] with representatives of the five countries concerned: Austria, Slovenia, Croatia, Hungary and Serbia. Many portions of the proposed territorial mosaic are already protected, in accordance to the respective national and local legislations; but a proper internationally recognized framework is still missing. The scenario of the *Trans-boundary Biosphere Reserve* is seen as 'flexible' enough – no rigid procedures are in place–and internationally recognized – even if not legally binding – to accommodate the existing initiatives while facilitating the next ones. Ultimately, it seems that the fact of scaling up the trans-boundary cooperation may help to find a solution to border issues that are still pending; for example, the one between Croatia and Serbia along the Danube.

4.2 The Dinaric Arc Initiative: A 'Model' Cooperation Platform

"I should speak now, but I am speechless. I regret that the rest of the world is not here to see what can be achieved in a region with heavy historic burden", said

---

[31] The workshop was held in Serbia, on July 24–26, 2009.

Mr. Tamas Marghescu, at that time Director of the IUCN *Regional Office for Europe*, on the occasion of his speech, given at the end of a very meaningful ceremony: that of the *Big Win for Dinaric Arc*, organized by the *Dinaric Arc Initiative* (DAI) partners on May 29, 2008, during the *9th Conference of the Parties* to the *Convention on Biological Diversity* (CBD 9COP) in Bonn (Germany). The cover picture of the WWF event report shows six ministers standing nearby each other, representing the high commitment of their respective countries–Albania, Bosnia and Herzegovina, Croatia, Montenegro, Serbia and Slovenia – to "foster a joint cooperation on conservation and sustainable development of the Dinaric Arc eco-region".[32]

This event well symbolizes the evolution of the DAI, which reflects a progressive path of concerted actions, initially[33] driven by a scientifically sound approach – the eco-regional one (see the Fig. 7) and later on transformed into an innovative inter-agency governance regime. Within its 5 years of existence, the Initiative' partners[34] have been constantly searching for areas and sites of common interest, as locations for joint actions to be promoted and implemented.

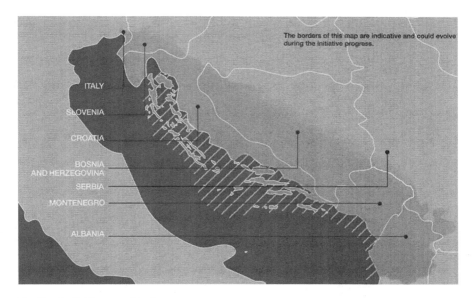

Fig. 7 The DAI geographical scope

---

[32] For the detailed report, see: www.panda.org.

[33] The DAI activities started with a first meeting held at WWF Med PO in Rome, in December 2004.

[34] Currently, DAI members are: UNESCO Venice Office, UNDP (country offices of Croatia, BiH, Montenegro and Albania), UNEP, FAO, Council of Europe, IUCN, WWF, ECNC, REC, Euronatur, SNV and ECNC.

In the most significant cases – the ones of the *Skadar/Shkodra Lake* and *Durmitor* and *Sutjeska National Parks* – the coral recognition from the partners of the importance of the UNESCO designations signifies the added value given to those territories by the international recognitions. The idea of establishing a *Trans-boundary Biosphere Reserve* encompassing the Montenegrin and the Albanian sides of the lake – in the first case – and the possibility to extend the existing '*Tara River Basin Biosphere Reserve*' across the border with Bosnia and Herzegovina to include the territories of the National Park '*Sutjeska*' – in the second case – are well recognized by the DAI partners – and their respective governmental counterparts – as the most appropriate 'territorial designations' to frame the existing activities and those to come.

5 Conclusion

The rapid *excursus* provided by this paper tends to demonstrate that "international organizations are better equipped in dealing with environmental security", as stated by Francioni.[35] The combination of 'constituency' and 'pragmatic' elements may certainly favor the effectiveness of most of the UN agencies work in fostering the environmental diplomacy.

Amongst the former, the ideological components – referring explicitly to their constitutions and mandates – as well as the political 'personality' gave authority to the international community action and capacity to 'upgrade' the negotiation among the parties, by 'scaling-up' from the local to the international level.

On the other hand, the capacity to pull intellectual resources and to activate well structured and nationally widespread operational mechanisms in combination with a standard setting 'toolkit' (primarily composed by conventions and treaties), made the multilateral system the most efficient international regime in place to foster environmental security.

Within this framework, the specific role of UNESCO and its designated territories offers additional chances; "the high number of countries that ratified the Convention is significant, as it gives it legitimacy and future" stated Cowie and Winbledon.[36] In fact, despite the relatively long history of the Convention, its 'vitality'[37] is still very high. If on the one hand, in agreement with Musitelli,[38] the growing number of designated sites cannot be considered the only significant indicator of 'success' of this treaty, on the other hand, the high political commitment that is given to its activities by the *State Parties* and the related stakeholders, makes it "one of the most successful examples of international co-operation".[39]

---

[35] On the occasion of the workshop 'International Regulation Implementation for Environmental Security in South-Eastern Europe', held at S. Servolo Island, Venice, December 3–5, 2009.

[36] Cowie and Wimbledon (1994), p 72.

[37] Dutt (1999) refers to the convention as one of the few "vital UNESCO instruments".

[38] Who wrote that "in order "to have a fair appreciation of the local impact of *World Heritage*, it is better to consider the sites themselves rather than the number of listings" (Musitelli 2003; p 335).

[39] Pocock (1997), p 268.

The contribution to the international cooperation and the environmental security given by the "ideas and ideals"[40] incorporated by the UNESCO designation process is still far from being completely exploited and their further use in the SEE strongly encouraged.

References

Castoriadis C (1984) Crossroads in the Labyrinth. Harvester, Brighton
CITIES (1973). Convention on International Trade in Endangered Species. Washington
CMS Secretariat (1979). Convention on the Conservation of Migratory Species of Wild Animals. Bonn, Conference of the Parties: 7
Cowie JW, Wimbledon WAP (1994) The world heritage list and its relevance to geology. In: O'Halloran D, Green C, Harley M, Stanley M, Knill J (eds) Geological and landscape conservation. Geological society, London, pp 71–73
Dutt S (1999) The role of intellectuals and non-governmental organizations in Britain's relations with UNESCO. The Round Table 88(350):207–228
EC (2010) Countries: enlargement. European Commission
Gaudry K-H (2007) Protected Areas' initiatives and Local Governments: Interpreting disourses on development. *Institute of Cultural Geography and Institute of Forest and Environmental Policy*. Master of Science "Environmental Governance". Freiburg, Albert-Ludwigs University of Freiburg, 89
Hösch E (2004) Geschichte des Balkans. C.H. Beck, München
Jones GJ (1994) World heritage sites. In: Unwin T (ed) Atlas of development. Wiley, Chichester, pp 316–318
Musitelli J (2003) World heritage, between universalism and globalization. International Journal of Cultural Property 11(2):323–336
Pocock D (1997) Some reflections on world heritage. Area 29(3):260–268
Rekacewicz P, Kluser S et al. (2007) Political history of the Balkan region. *Balkan Vital Graphics*. U. G.-A. M. a. G. Library, UNEP/GRID-Arendal
ReliefWeb (2000) UN OCHA in South Eastern Europe. UN OCH Affairs
SEE-TCP (2010) Programme area, South Eastern Europe. S. E. E.-T. C. Programme
Swoboda H, Solioz C (eds) (2007) Conflict and Renewal: Europe Transformed. Essay in Honour of Wolfgang Petritsch. Nomos, Baden-baden
UNESCO (1971). *Final Act of the International Conference on the Conservation of Wetlands and Waterfowl*. Ramsar, Iran, International Wildfowl Research Bureau
UNESCO (1972) Convention concerning the protection of the world cultural and natural heritage adopted by the general conference at its seventeenth session, Paris, 16 November 1972 at: http://whc.unesco.org/en/conventiontext
UNESCO (1994). Convention on Wetlands of International Importance especially as Waterfowl Habitat. The Convention on Wetlands text, as amended in 1982 and 1987. O. o. I. S. a. L. Affairs. Ramsar, Ramsar Convention Bureau
UNESCO (2007) UNESCO biosphere reserves: model regions with a global reputation. UNESCO Today 2
UNESCO (2008) Operational guidelines for the implementation of the world Heritage Convention at: http://whc.unesco.org/en/guidelines/
United Nations U (1993). Multilateral convention on biological diversity (with annexes). Concluded at Rio de Janeiro on June 1992. Rio de Janeiro, United Nations – Treaty Series. 1760

---

[40] This was the 'definition' of the *Biosphere Reserves* given by Bridgewater – former Director of the UNESCO *Division of Earth and Ecological Science* – on the occasion of the *World Conference on Biosphere Reserve*, held in Madrid, February 2008.

Social and Environmental Issues Related to Security in SEE Countries

Dragoljub Todić

Abstract In this paper different and very complex methodological problems concerning understanding of the relation between social and environmental issues on the one side and security risks on the other are emphasized. Several important general social issues characteristic for countries of SEE (geographic position, level of economic development and cooperation, historical heritage, state of awareness of the public concerning specific social questions, relation with the EU) as basic factors which influence understanding of general frameworks of security in this region have been underlined. A special attention is paid to the state of the environment and its relevance for security in SEE. Within this context, an assessment is provided and critical elements in the fields of nature protection, biodiversity, waste management, water management, agriculture, energy, transport, harmonization of national regulations with those of the EU are pinpointed. The central thesis is that the specific combinations of social, economic and environmental characteristics of countries of the region are causing (or can cause) serious instability and threaten security. The low level of mutual cooperation between the countries of the region is another particularly unfavourable circumstance influencing security of the region.

Keywords Balkan • South East Europe • Security • Environment • Economic cooperation • Migrations • Energy • Climate change • Waste • Nature • Water • Transport • Harmonisation with EU • Multilateral agreements • Legal harmonisation

1 Introduction

In a more precise analysis of the interconnection of social and environmental issues with security issues in SEE, several methodological questions deserve special attention. The first of these questions relates to the establishment of criteria for measurement

D. Todić (✉)
Graduate School of International Economy, Megatrend University, Belgrade, Serbia
e-mail: dragoljubtod@yahoo.com

(quantification) of the "relevance" of specific social and environmental issues for understanding the notion of "security." It leads us basically to a preliminary definition of the essence, i.e. contents of the notion of "security" which is (or which could be) of importance for "social and environmental issues" in SEE.[1] (Vuori 2001, p. 26.) A preliminary precise definition of the subject of protection (i.e., "subject of endangerment" and "protected values"), would probably contribute to clarify the analysis and conclusions. If "human security,"[2] i.e. "survival of humanity" as a universal value and other related values (human health, human rights, economic development, and so on), are taken for the indisputable subject of protection, then the list of issues that could be relevant for the discussion is very widely defined. A strong interaction of the causes related to specific wider socio-security phenomena and their consequences contributes to the complexity of the notion. In this sense, the security risks of the individual member countries of the region exacerbate the methodological complexity of the "international security" issue. Furthermore, the question of the relation between the classic forms of crime (particularly organized crime) (Xenakis 2004), and some environmental problems should be added to this, although "the level of organized crime groups' involvement in environmental crime is difficult to determine, mainly because it has been a low priority for law enforcement authorities in some Member States…." (EU 2005).

In connection with this, there is also the question of how much and in which way certain social issues, i.e. environmental issues, find their expression in the "national interest" of states as subjects of international relations and international law. This is especially true if we have in mind the contemporary global trends in environmental policy. The intermingling of global, regional and national dimensions of contemporary social and environmental problems and their impact on different aspects of security is one of the most evident characteristics of contemporary world. Because of that, it should be added the interesting question of how much and in which way these issues are intermingled as security policy issues of individual states in the region and of international organizations to which they are members. The universalization of values, instruments and goals, in contemporary environmental policy only underlines bilateral and multilateral cooperation of states in the region as an imperative dimension and as a general framework for understanding security as a whole (Todić 2008).

If certain social issues of a general character are to be considered in a relatively simple way (as their framework character, preconditions etc.), it seems that a more precise methodology is needed for the exploration of the connections between

---

[1] South-eastern Europe: Albania, Bosnia and Herzegovina, Croatia, Former Yugoslav Republic of Macedonia, Serbia, Montenegro; Other SEE countries: Bulgaria, Romania, Turkey. (European Environment Agency (2007), p. 21.) See also, UNDP 2007b, p. 7. For purposes of this paper most often is taken into account or analyzed the state of the group of SEE countries belonging to Western Balkans, except in cases when explicitly is mentioned something else. For some dilemmas concerning the origin of the word "Balkans" and controversies concerning the notion and its use (for example "balkanization"), see Gleni 2001, pp. 11–14; Cvijić 1922, pp. 1–8.

[2] "Human security" refers to a life in peace, without tension and conflict – political, social or economic – between people and nations. It encompasses freedom from a range of risks: disasters, hunger, diseases, civil strife or war, terrorism, eviction, persecution or discrimination, financial deprivation and poverty, and more (ECE 2007d, p. 3.). See also: O'Brien and Leichenko 2007, p. 3–7.

environmental issues and security risks (Snoy and Baltes 2009, p. 314). Beside the questions linked to the definition of the categorical apparatus, it seems that the basic problem is the one of understanding the relation between security and the environment mostly in the sphere of the realistic possibilities of theoretical and practical extrapolation of environmental segments from the totality of social processes as a whole. (Vukasović 1997) It is obvious that some aspects of the methodological problems relate to the very broad legal definitions of the notions of environment and other relevant formulations in the national legal systems as well as in international environmental law (Todić 1999). This contributes significantly to uncertainty between the specific characteristics of the environmental problems in relation to other relevant issues and their impact on the "state" of the security. Because of that, the possibility for precise elaboration of these issues is much more complex though "in many instances, the environmental/conflict linkages are readily apparent" (Myers 2009, p. 47).[3]

2 Social Issues Related to Security in SEE Countries

2.1 General Social Characteristics

Several general common social and economic characteristics of the countries in the region, which are directly relevant (or could be relevant) for understanding contemporary security dilemmas, derive as a result of various factors including the following: geographical (and geopolitical)[4] position of countries and the region as a whole (Stojković 1997), post-communist and transitional character of their economic and political systems developed in the last 20 years, overall and very complex historical heritage and inter relations between states of the region (Bechev 2004),[5] etc.[6] This region is highly diverse in terms of its ethnic groups (Sugar and Lederer 1994), religions,[7] culture, ecosystems,[8] etc.

Generally speaking, countries of the SEE region share a common economic and geopolitical background, insofar most of them having previously been part of the

---

[3] Hence the definition of the term of environmental crime seems very broadly ("any breach of a national or international environmental law or treaty that exists to ensure the conservation and sustainability of the world's environment, biodiversity, or natural resources.") http://www.interpol.int/Public/ICPO/FactSheets/PST03.pdf

[4] Neighbors to the European Union and North Atlantic Treaty Organization; a bridge to the Middle East and Central Asia, etc.

[5] Across the region, historic animosities (and national politicians) were said to be the major obstacles that kept the countries of the region from growing closer together (Gallup Balkan Monitor Insights and Perceptions: Voices of the Bal-kans – 2009 Summary of Findings, p. 9).

[6] Quite concrete connections between these questions and security in SEE countries should be determined case by case, what is beyond the scope of this paper, in which we are considering principal issues.

[7] Because of its location at the boundaries between past empires and cultures the major religions in region are Catholic, Orthodox and Islam.

[8] Four different Europe's bio-geographical areas: Central European, Mediterranean, Pannonic and Alpine.

former Yugoslavia. Since 1990, the economic, social and environmental risks to security in Europe have been, to a large extent, associated with the uncertainty about the outcome of the process of transition towards market economy systems and multi-party political democracies in the countries of Central and Eastern Europe (UNECE 2003, p. 5). The situation in this part of Europe was severely disturbed during instability and armed conflicts (1991–2000) that followed the dissolution of the Socialist Federal Republic of Yugoslavia and collapse of the former (socialist) political systems, wars[9] and the NATO intervention (UNEP 2003a; Economic Commission for Europe 1999, p. 6; UNEP 1999; UNEP 2002; UNEP 2004b; Federal Ministry for Development, science and the environment 2000, pp. 5, 7, 96, 106). The conflict in Yugoslavia erupted at a time when no international organization was willing or prepared to take the lead in managing what became a particularly protracted intra-state conflict (Stewart 2004, p. 232). Although an assessment of the war's consequences for some countries in the region has been conducted, it seems that an integrated and reliable estimation of the overall consequences of the war in the region has not been produced yet. However, several (social, economic, environmental, political, etc.) issues emerged on the ground of the so called Yugoslav crisis (related to Kosovo under UNSCR 1244/99, Republic of Srpska and Dayton agreement) could be considered as extremely challenging.

All the SEE countries experienced major internal and external migrations during the last two decades because of the war and the effects of economic restructuring. Negative demographic consequences caused by the conflicts will probably never be completely remedied. They have an impact on the state of the environment and cause, among other consequences, further polarisation of the level of economic and social development between the urban and rural areas in the region and between different parts of the region (UNDP 2007, p. 168; European Environment Agency 2007b, p. 26).

2.2 *Basic Economic Characteristics and Mutual Cooperation*

Dominant economic problems in the region remain serious and include high unemployment, significant and rising trade imbalances, increasing external debt, growing discrepancies in economical power among regions and citizens. The State intervention in the economy is significant and still includes examples of costly subsidies to loss making enterprises. Apart from that, it could be said that the economies of the region are "small economies" (European Environment Agency 2007, p. 390), the volume of mutual trade and cooperation is small[10] and the level of trade with the EU is high (Koyama 2003, pp. 185–187), the level of harmonization with the EU and international standards is unsatisfactory, the level of competitiveness is unsatisfactory, the

---

[9]Therefore, needs for continued monitoring and involvement of the international community at several levels – political, economic and military ones – are suggested (Bilandžić 2001, p. 57).

[10]Notwithstanding, it seems that this fact has not been highlighted in relevant Environmental performance reviews. See Economic Commission for Europe 2002a, pp. 33–44; Economic Commission for Europe 2004, pp. 55–68; EC 1999, pp. 37–59; Economic Commission for Europe 2002b, pp. 45–54; Economic Commission for Europe 2007, pp. 45–58.

level of "grey" economy is high. The proportions of the population living below the poverty line are still significant in some parts of the Balkans and the inter-connection between poverty and environmental degradation has been proved in different fields, including environmental security (UNDP 2007b, pp. 23–24). Majorities in all countries (except in Albania and Kosovo under UNSCR 1244/99) felt that their country was going in a bad direction and at least half of respondents expected their standard of living to worsen in the years to come (Gallup Balkan Monitor Insights and Perceptions: Voices of the Balkans – 2009 Summary of Findings, p. 7).

All the countries in the region have undergone or are still undergoing the processes of privatisation that has a fundamental effect, inter alia on the application of the principle of responsibility for the environment and sustainable development based on the "polluter pays" principle. There is a general public perception of corruption in privatisation deals leading to mistrust in the environmental performance of privatised companies.[11] In general, with regard to the good governance and corruption more than two-thirds of Balkan respondents thought that corruption was pervasive in business and government (Gallup Balkan Monitor Insights and Perceptions: Voices of the Balkans – 2009 Summary of Findings, p. 8).[12]

Another problem which should be mentioned is the fact that the economies of the region have been largely based on agriculture, natural resources and industry which caused a high rate of resource depletion and environmental pollution.[13]

Regarding economic cooperation between the countries in the region, the Agreement on Amendment of and Accession to the Central European Free Trade Agreement (CEFTA 2006) signed under the auspices of the Stability Pact for South Eastern Europe has a special significance.[14]

2.3 Relations with EU, NATO and Other Relevant Organisations

The relevance of relations between the countries of the region and the EU stems from the fact that the total multidimensional relations have (or could have) different security components (Stewart 2004, pp. 223–248).

---

[11] Moreover, there is a "generally negative attitude towards private sector service providers linked to a perception that they make profit out of environmental problems. Perceptions of corruption are also a deterrent to contracting private companies" (UNDP 2007b, p. 78).

[12] According to the Corruption Perceptions Index 2008 Croatia is 62nd on the list, Romania is 71st, Bulgaria is 72nd, Macedonia is 74th, Albania is 85th, Montenegro is 88th, Serbia is 91st, Bosnia and Herzegovina 93rd. pp. 397–402. On this see also: Corporate Author Transparency International 2009, pp. 409–412.)

[13] Notwithstanding this, it is believed that the main competitive advantages of the countries in the region are natural resources (forests, high biodiversity, minerals, coal, water, etc.) apart from their human resources (a relatively high level of education) (UNDP 2007b, p. 19).

[14] The Agreement entered into force on 26 July 2007 for five Parties – Albania, Macedonia, Moldova, Montenegro and UNMIK/Kosovo, while for Croatia it entered into force on 22 August, for Serbia on 24 October and for Bosnia and Herzegovina on 22 November 2007. http://www.cefta2006.com/en-index.php

Although several barriers to the EU accession process have been identified, generally speaking EU accession is the principal objective of all countries and other territories in SEE.[15] Generally speaking, public support for EU accession has remained high in all countries of the region except Croatia (Gallup Balkan Monitor Insights and Perceptions: Voices of the Balkans – 2009 Summary of Findings, p. 7). Different countries and other territories are in a different phase of the accession process. In the 2009 European Community' Commission Progress Reports, the relations between countries and territories of the SEE region and EU have been described as follow: Albania is a potential candidate for EU membership (SEC 2009a, pp. 4, 5); Bosnia and Herzegovina is a potential candidate for EU membership (SEC 2009b, p. 4); Montenegro is a potential candidate for EU membership (SEC 2009f, pp. 4, 5); Serbia is a potential candidate for EU membership[16]; Kosovo under UNSCR 1244/99 is participating in the Stabilisation and Association Process.[17] The European Council of June 2004 granted the status of candidate country to Croatia (SEC 2009b, p. 4); and the European Council of December 2005 granted the status of candidate country to the Former Yugoslav Republic of Macedonia (SEC 2009c, p. 5).

Relations of SEE countries with NATO can be interpreted with different arguments (Angelov 2004, p. 46) although the "neutral" position of Serbia[18] could be considered as some kind of deviation from general tendency to become member of the NATO.

3 Environmental Issues Related to the Security in SEE Countries

3.1 State of the Environment

Although different documents, scientific papers and reports underline different specific problems in the region, generally speaking, "environmental quality in SEE is roughly in line with global averages, albeit well below EU levels." Nevertheless, the national MDGs reports for SEE states emphasize air and water pollution, as well as protecting areas devoted to forests and biodiversity. (UNECE 2003, p. 49) There are a wide range of environmental pressures (biodiversity loss, contamination of water

---

[15] At the same time all SEE countries are member of the Council of Europe and the Organization for Security and Co-operation in Europe.

[16] Serbia has started to implement provisions of the Interim Agreement. Implementation by the EU countries has been conditioned by the cooperation with International Criminal Tribunal for the former Yugoslavia (ICTY). (SEC (2009g, pp. 4–5)).

[17] In February 2009, the European Parliament adopted a resolution underlining its willingness to assist the economic and political development of Kosovo by offering clear prospects for EU membership (SEC 2009d, p. 5).

[18] See the Resolution of the National Assembly of the Republic of Serbia, paragraph 6 (26 December 2007); http://www.parlament.gov.rs/content/cir/akta/aktadetalji.asp?Id=360&t=O# Accessed 20 October 2009.

and soil, illegal or inappropriate disposal or treatment of hazardous and conventional waste, legal or illegal dumping of depleted uranium, unsustainable agricultural practices and so on) (UNEP 2003b) with significant impacts on human health[19] and on economic development. The region is mainly affected by heavy industrial pollution in urban-industrial areas, intensive agriculture with yet uncalculated health impacts, lack of water technology and infrastructure, and industrial pollution from the mining sector (UNEP 2003b, p. 16). The inappropriate disposal and treatment of waste from municipal and industrial sources, and regular industrial activities, including inadequate storage of chemicals, are the major sources of contamination (European Environment Agency 2007, p. 118). Moreover, SEE countries are exposed to a range of disasters caused by the impact of natural hazards, including earthquakes, floods, forest fires, drought and landslides (World Bank 2008).

3.1.1 Nature and Biodiversity Conservation

The economic development in these countries has brought a number of threats to biodiversity including the following: overexploitation of resources due to poverty, intensification of agricultural and forestry practices, urbanisation, land abandonment, changes in water regime due to construction of dams and irrigation, pollution and so on. Rivers, coastal zones, and wetlands are the most threatened in the short term, and in the longer term the mountain meadow ecosystems are also considered vulnerable (UNDP 2007b, p. 119).[20]

There is a lack of relevant information about the quality and quantity of biodiversity in the region; there is a lack of effective inter-sectoral cooperation in the field of protection of biodiversity, and a lack of integration of biodiversity protection into sectoral development policies, and so on. Implementation of physical and urban plans, construction permits, as well as management system for collection and trading of wild flora and fauna, are in many cases inadequate to meet the conservation objectives. Management systems and resources for protected areas are still developing, and where they are insufficient, poor enforcement and negative trends which endanger some of the important biodiversity values in the region are recorded (UNDP 2007, pp. 125–126; Dimoski and Glaser 2002). Institutional and methodological discontinuity causes a lack of up-to-date forest inventories and of reliable data concerning the status of forest health in some countries. Integration of environmental

---

[19] "The main environmental health hazards in the region are linked to: air pollution (both outdoor pollution from major industries, mining and transport, as well as indoor pollution from e.g. smoking, heating); incidence of unsafe drinking water in certain areas/time periods; improper management of waste and wastewater; and insufficient occupational (in particular in major industries) and transport safety." (UNDP 2007b, p. 155). See extensively: Economic Commission for Europe 2002a, pp. 145–160; Economic Commission for Europe 2004, pp. 149–162; EC 1999, pp. 169–176; Economic Commission for Europe 2002b, pp. 165–184; Economic Commission for Europe 2003, pp. 203–216).

[20] See also: Economic Commission for Europe 2002a, pp. 99–112; Economic Commission for Europe 2004, pp. 99–112; Economic Commission for Europe 2002b, pp. 109–120; Economic Commission for Europe 2003, pp. 117–132.

and nature conservation objectives into forest planning and practice is also negatively influenced by separated institutional responsibilities (UNDP 2007b, p. 146).

Co-operation in managing nature and biodiversity in the context of integrated sustainable development should be further strengthened on a trans-boundary level. A significant number of countries in the region participate in the implementation of existing multilateral environmental agreements in the field of nature conservation and biodiversity. However, further progress could be achieved through the proposed SEE Mountains Initiative and Framework Convention on the Protection and Sustainable Development of South-Eastern Europe Mountain Regions, building upon experiences learnt from implementation of the Alpine and Carpathian Conventions (ECE 2007c).

3.1.2 Availability and Water Consumption

One of the major resources of SEE is water, managed by different sectors and authorities in each country and territory (UNDP 2007b, p. 127; Economic Commission for Europe 2002a, pp. 69–86; Economic Commission for Europe 2004, pp. 89–98; Economic Commission for Europe 1999, pp. 69–88; SEC 2009e, pp. 55–74; Economic Commission for Europe 2007, pp. 81–94; Federal Ministry for Development, science and the environment 2000, pp. 61–80). It is a pity that in most countries water monitoring is not performed according to the programmes and legal requirements. The information collected is not compatible with the EU monitoring requirements and none of the Core Set of Indicators related to water are readily available. The countries have started to work on the EU Framework Water Directive implementation including river basin management. The progress has been, however, slowed down by, among other causes, institutional complexity, conflicting interests regarding different water uses and a lack of efficient mechanisms to resolve them. Furthermore, there are problems with access to existing information. A number of institutions are competent for water in each country, which causes a lack of consistency in water management and frequent institutional conflicts regarding specific issues. There is a chronic under-investment in water related infrastructure that is further worsened by the fragmentation and relatively small size of utility companies governed by individual municipalities (UNDP 2007b, pp. 132, 133).

With regard to the membership in MEAs and the implementation of existing MEAs in the water field, there is room for significant improvements. Bilateral regulation of relations among states of the regions that have common water resources is mostly not based on modern principles of water resources management.[21]

3.1.3 Marine Environment

The Adriatic countries participate in the UNEP's Mediterranean Action Plan and are parties to the Barcelona Convention on the protection of the Mediterranean or in the process of becoming members. In accordance with the Barcelona Convention Protocol

---

[21] For more details on state in Serbia, see Todić and Vukasović 2009.

Concerning Cooperation in Preventing Pollution from Ships and, in Cases of Emergency, Combating Pollution of the Mediterranean Sea, Croatia, Italy and Slovenia prepared, in 2005, a sub-regional contingency plan for the Adriatic Sea (UNDP 2007b, p. 138).

Monitoring of various parameters of the sea is not conducted in a comprehensive and continuous manner, and the data necessary for integrated management is only partially available. Understanding about the general status of the marine ecosystem is still insufficient. (Except in Croatia, UNDP 2007b, p. 139).

Differences in the level of economic development and individual country strategies complicate strategic vision, leadership and framework for sustainable use and preservation of marine ecosystems of the Adriatic. This makes the international negotiations about the fair use of marine resources and protection of the Adriatic quite difficult.

Attempts to achieve consensus about sustainable management of the marine resources of the Adriatic Sea are hindered by many vested interests in the fisheries sector. Only in Croatia there is a part of the Adriatic marine ecosystem currently protected, in contrast to the significant share of land area under protection in all the countries. Negative impacts of land based pollution are to be mitigated by the development of the necessary sewer and wastewater treatment infrastructure (UNDP 2007b, p. 139).

3.1.4 Climate Change

Generally speaking, climate change is seen as a serious international problem and not only as an issue of domestic character. (UNDP 2007b, p. 173). Negative consequences of the climate change are already causing droughts, flooding and increasing frequency of forest fires (UNDP 2007b, p. 170). Preliminary results of climate research and potential climate changes on the territory of Serbia, caused by global climate changes, indicate trends of further increasing of the air temperature and a considerable precipitation decrease, especially in the warm season of the year (ECE 2007b, p. 2).

It is likely that the security consequences of climate change should be viewed in the context of specific assessment of possible developments at the global level (total economic activity, for each industry – especially agriculture, migrations, and so on) (UNDP 2007b, pp. 33, 58, 243, 257, 318–322).

Although the absolute and per capita contribution of the countries in the region to global greenhouse gas emissions is relatively small, there are possibilities to improve emission reductions in the countries. This is due to the current high energy intensity, reliance on brown coal for energy production, low energy efficiency of the domestic sector and high potential for renewable resources such as solar, biomass, hydro and wind. After the end of war in the territory of the former Yugoslavia, transport has increased rapidly and it is becoming one of the main sources of pollution and emissions of CO_2. Road transport is increasing following also the rapid development of road networks, while public transport and railways are declining. (UNDP 2007b, p. 189).

One of the existing problems relate to the lack of data and accurate information on GHG emissions (particularly in Serbia, Montenegro, BiH, and Kosovo under UNSCR 1244/99). Apart from that, there are also very few long-term climate data series in the region and no analysis on vulnerability and adaptation strategies in some countries. There is also an important problem of institutional upgrading and coordination between different sectors (energy, industry, transport, agriculture, forestry, and others). As a result, there is no coherent foundation on which to integrate specific climate change mitigation and adaptation actions.

All countries in the region are parties to the UN Framework Convention on Climate Change of 1992 and the Kyoto Protocol.[22] Croatia belongs to the group of countries in transition (Annex I of UNFCCC) and the remaining countries are in the group of Non-annex I countries. The region could benefit significantly from emissions trading (Croatia) and the Clean Development Mechanism.

During the Belgrade Conference "Environment for Europe" (2007) the Belgrade Initiative to enhance SEE sub-regional cooperation by employing the mechanism of sub-regional planning through the SEE/CCFAP was launched (ECE 2007b, p. 24).

3.2 Sectors That Drive Environmental Changes[23]

3.2.1 Industry

Large polluting industries in mineral extraction, metal processing and energy production are specific for the region. South Eastern Europe showed "great" concern over the adverse impact of the mining practices (ECE 2007d, pp. 6, 34).

Companies often use obsolete technologies and are overstaffed, and although they are not attractive in business terms, for the governments it has been an imperative to maintain their operation for saving as many jobs as possible. The Baia Mare industrial accident in Romania in 2000, with severe transboundary pollution, demonstrated that operations involving hazardous substances are a serious threat to the environment, and showed that accidental water pollution can have far-reaching transboundary effects even if it occurs far from an international border (REC 2001).

It is positive that, environmental permits and self-monitoring are being introduced gradually through the definition of framework environmental legislation and the more specific transposition of the IPPC directive and EIA procedure.

The transposition of the EU *acquis* also includes EMAS and eco-labels, which were introduced quite early in some of the countries, but with limited success due to the small size of the markets and the lack of domestic consumer demand (UNDP 2007b, p. 184). The number of the ISO 14001 certifications (from 2005 to 2007) in the SEE region slowly increased but it is still below European level (ISO 2009, p. 26).

---

[22] http://unfccc.int/files/essential_background/convention/status_of_ratification/application/pdf/unfccc_ratification_20091016.pdf http://unfccc.int/files/kyoto_ protocol/status_of_ratification/application/pdf/kp_ratification_20090826corr.pdf; Although with undefined final status, Kosovo, UN administered territory under UN Security Council Resolution 1244 – through UNMIK, is an equal signatory of the "Energy Community Treaty for the SEE Countries," and is obliged to begin the implementation of the Kyoto Protocol.

[23] The three sectors focused on here are not the only ones that need to be considered. Tourism, transport, households are also important sectors with respect to environmental impacts.

3.2.2 Energy

A lot of energy in SEE countries comes from coal and fuel wood is still a very important heating source. In the context of concerns such as energy security, adequate access to energy services, particularly in rural areas, and modernisation of energy infrastructure, it is unclear whether the SEE will be able to reduce its energy intensity in the near future (UNDP 2007b, p. 175; Economic Commission for Europe 2004, pp. 135–148; Economic Commission for Europe 2002b, pp. 121–138; SEC 2009d, pp. 85–94; Economic Commission for Europe 2007, pp. 95–110).

The advantage of the region in terms of low energy use per capita (which is found in most countries) will be lost if current development patterns are continued. One of the problems relates to the capacity to initiate, support and implement efficiency and renewable projects (which is still limited). Apart from that, awareness and information on available energy efficiency and renewable technologies and measures and their financial benefits is scarce (European Environment Agency 2007, p. 404). For investment in small-scale energy efficiency and projects to exploit renewable energy sources, only limited funds are available.

As a significant contribution in the energy sector we can considered the conclusion of the Treaty establishing the Energy Community, signed 25 October 2005 in Athens by the European Community and then nine Contracting Parties from South East Europe.[24] It entered into force on 1 July 2006. This treaty relates to question of security in several different ways, through provisions that regulate various aspects connected to security, as well as trough provisions explicitly devoted to security.[25] The majority of countries in the region (not all) ratified the Energy Charter Treaty and Protocol on Energy Efficiency and Related Environmental Aspects.

Cooperation with Russia in the energy field (i.e. building regional oil and gas pipelines) has strategic security and environmental importance.

Having in mind the existence of nuclear power plants in some countries of the SEE region (and in neighbouring countries) controversy on security risks could get new dimensions.

3.2.3 Agriculture

Given the fact that traditional agriculture in SEE countries is undergoing significant changes, there are different influences on the state of the environment. Despite good natural conditions for farming and a high rate of unemployment, a lot of land has been abandoned (European Environment Agency 2007, p. 295). On the other hand, there has been further intensification in farming and food processing generating

---

[24] Albania, Bulgaria, Bosnia and Herzegovina, Croatia, FYR of Macedonia, Montenegro, Romania, Serbia and The United Nations Interim Administration Mission in Kosovo pursuant to the UNSCR 1244/99. See: http://www.energy-commnity.org/portal/page/portal/ENC_HOME/ENERGY_COMMUNITY/Legal/Treaty

[25] Due to that, one should take into account when construing this treaty, its provisions as a whole, including the provisions related to the EU *acquis on* environmental law.

negative impact on environment. In some countries a significant share of the farmland is contaminated with land mines (UNDP 2007b, pp. 112–118). The main sources of soil pollution are animal farms, especially large-scale pig and poultry breeding units. The treatment of fertilizers and manure still does not appear to be environmentally optimal. Apart from that, there is a significant environmental legacy where localized hot spots of contamination are often associated with the storage and disposal of pesticides (European Environment Agency 2007, p. 297).[26]

3.3 Participation in MEAs

Due to the fact that most multilateral environmental agreements contain provisions relating to different security issues, at least indirectly, MEAs have played an important role in the process of environmental improvement in the region, in domestic and international co-operation,[27] on issues of sustainable development and in the overall process of democratisation.

Most of the global MEAs have been ratified by the countries of the SEE region (UNDP 2007b, p. 206). In case of regional UNECE conventions Albania and Croatia are party to all of them, unlike the other countries in the region (ECE 2007a, pp. 4, 21, 367–369). It may be disputable that the Convention on Transboundary Effects of the Industrial Accidents is not generally accepted in the region as well as the Convention on the protection and use of Transboundary Watercourses and International Lakes. Nevertheless, the relevant provisions implementation could be considered as an obstacle. There is a need for further improvement and enhancement of the administrative capacity to address environmental problems in the SEE region (ECE 2007a, p. 4).

Having in mind all the above mentioned common problems, the establishment of regional framework agreements on environmental and security cooperation could be considered.

3.4 Harmonisation of National Environmental Legislation with the EU Legislation

In the process of harmonisation of national legislation with the EU legislation several common problems have been identified (institutional capacities, institutional division of the responsibilities, awareness, enforcement) (Dimoski and Glaser 2002; UNDP 2007b, pp. 53–54; REC 2006, pp. 31–52). In 2009, the Albanian implementation and

---

[26] For more details see Economic Commission for Europe 2002a, pp. 121–132; Economic Commission for Europe 2004, pp. 121–134; Economic Commission for Europe 2002b, pp. 95–108; Economic Commission for Europe 2003, pp. 165–176.

[27] See more extensively see Economic Commission for Europe 2002a, pp. 33–44; Economic Commission for Europe 2004, pp. 55–68; Economic Commission for Europe 2002b, pp. 45–54; 39–52; 36. Economic Commission for Europe 2007, pp. 45–58.

enforcement remain weak. Preparations in the field of environment are advancing slowly but remain at an early stage for a number of sectors (SEC 2009a, pp. 37–38). Bosnia and Herzegovina, for instance, needs to strengthen its environmental protection institutions, in particular at State level. It also needs to mainstream environmental concerns in other sectors and to step up its efforts with regard to its obligations under international conventions (SEC 2009b, pp. 47, 48). In Croatia, considerable efforts are still needed in the water sector and nature protection, especially with regard to the implementation of the Water Framework Directive and the designation of Natura 2000 sites. Implementation of the horizontal *acquis* needs to be improved. Administrative capacity needs further strengthening both at national and at local level (SEC 2009c, pp. 61–63). In Montenegro implementation and enforcement need to be further strengthened. Particular attention needs to be paid to strengthen administrative capacity and to establish effective inspection services. Work on awareness-raising on environmental protection issues needs to be continued. Environmental protection, in particular in coastal areas and national parks, remains a cause for concern (SEC 2009f, pp. 38–40). In Macedonia, some sectors, like water quality or IPPC, are still lagging behind. Implementation of the legislation remains a considerable challenge. Administrative capacity is weak at both national and local levels (SEC 2009e, pp. 68–70). In Serbia enforcement of the legislation needs to be improved at all levels. In this regard, further efforts are needed to build up administrative capacity (SEC 2009g, pp. 43–45). Kosovo under UNSCR 1244/99 is at an early stage in alignment with European environmental standards (SEC 2009d, pp. 37, 38).

4 Conclusion

1. The need for defining more precise connections between social, economic and other characteristics of the region of SEE, or Western Balkans, on the one side and security risks on the other, creates the necessity of establishing preliminary reliable methodological tools. This is similar with the ascertainment and analysis of the connections between the state of the environment in the region and the security challenges stemming from specific domains of the environment (waste management, water management, energy), which are relatively easy to identify and measure in their impact (risks) on human health and environmental security.
2. On the general discussion level, it can be concluded that several common characteristics of the states of the region may give rise to security risks (level of economic development, percentage of unemployment, historical heritage, geostrategic position) although all of them at the same time represent open opportunities for possible improvement of cooperation. The lack of mutual cooperation and the absence of adequate instruments of cooperation can be considered as one of the most serious problems with possible security implications.
3. On the other side, in the part related to the state of the environment and the state of environmental governance in the countries of this region, it could be discussed in more concrete terms about open questions, weaknesses of the management system

and possible security risks in the region. The improvement of the state of the environment and the attainment of desirable level of management in different fields (upgrading of the level of energy efficiency, waste management, nature protection and protection of biodiversity, water management, transport) could significantly contribute to improve the security level in the region, even without a precise definition of the notion "security." Probably, management of common resources could be regarded as one of most delicate questions, the improvement of which should be accepted as one of the priorities of cooperation for the states in the region.
4. The strengthening of mutual cooperation of the states in the region within the framework of existing international (global and regional) legal instruments is good, but the development of specific instruments of cooperation could also contribute to more successful solving of dilemmas and open questions. The drafting and conclusion of regional multilateral agreement on environmental cooperation could be one of the most interesting ideas. The existing initiatives for the establishment of new agreements and arrangements (mountain, climate change, and so on) should be supported.
5. Since the states of SEE region are in the process of association with the European Union, solving security challenges related to the environment can be achieved through the improvement of specific elements of the environmental management system. The framework consists in the harmonisation of the national legal regulation with the EU legal regulations and in the development of institutional capacities in general. Therefore, the slowdown in the European integration process of this region as a whole or of particular countries could become a security risk. On the contrary, the speed-up of this process could create a basis for fundamentally improving regional stability.

References

Angelov A (2004) NATO enlargement and the Balkan. In: Fatić A (ed) Security in Southeastern Europe. The Management Centre, Belgrade, pp 99–120

Bechev D (2004) Worst-case scenarious and historical analogies: interpreting Balkan interstate relations in the 1990s. In: Fatić A (ed) Security in South-eastern Europe. The Management Center, Belgrade, pp 29–52

Bilandžić V (2001) Perspektive regionalne stabilnosti na Balkanu. (Prospects for Regional Stability in the Balkans). Int Prob 1(2):40–58

Corporate Author Transparency International (2009) Global Corruption Report 2009, Corruption and the Private Sector, Transparency International. Cambridge University Press, Ernest & Young

Council of the European Union (2005). EU Organized Crime Report 2005, Public Version, 13788/1/05, Rev. 1. Hague

Cvijić J (1922) Balkansko poluostrvo i južnoslovenske zemlje, osnove antropogeografije (Balkan Peninsula and South Slav countries, basics of anthropogeography. Državna štamparija Kraljevine Srba, Hrvata i Slovenaca, Beograd

Dimoski M, Glaser R (2002) Environmental enforcement and compliance in South Eastern Europe. The Regional Environmental Centre for Central and Eastern Europe, Szentendre

ECE (2007a) Acceptance and implementation of UNECE multilateral environmental agreements in South-Eastern Europe. ECE/BELGRADE.CONF/2007/INF/19

ECE (2007b) Belgrade initiative: enhancing the regional SEE cooperation in the field of climate change – climate change framework action plan for the SEE region, and the establishment of a sub-regional, virtual climate change related centre for research and systematic observation, education, training, public awareness, and capacity building. ECE/BELGRADE.CONF/2007/20

ECE (2007c) Environment and security partnerships: conflict and the environment. ECE/BELGRADE.CONF/2007/26

ECE (2007d) Protection and sustainable development of mountain areas in South Eastern. ECE/BELGRADE.CONF/2007/INF/23

Economic Commission for Europe (1999) Croatia – environmental performance review. Economic Commission for Europe, Committee on Environmental Policy, New York and Geneva

Economic Commission for Europe (2002a) Albania – environmental performance review. Economic Commission for Europe, Committee on Environmental Policy, New York and Geneva

Economic Commission for Europe (2002b) Macedonia FYR – environmental performance review. Economic Commission for Europe, Committee on Environmental Policy, New York and Geneva

Economic Commission for Europe (2003) Yugoslavia – environmental performance review (2003). Economic Commission for Europe, Committee on Environmental Policy, New York and Geneva

Economic Commission for Europe (2004) Bosnia and Herzegovina – environmental performance review. Economic Commission for Europe, Committee on Environmental Policy, New York and Geneva

Economic Commission for Europe (2007) Serbia – environmental performance review. Economic Commission for Europe, Committee on Environmental Policy, New York and Geneva

European Environment Agency (2007) Europe's Environment, Forth Assessment. European Environment Agency, Copenhagen

Federal Ministry for Development, science and the environment (2000) Yugoslavia – The Consequences of NATO bombing for the Environment in FR Yugoslavia (2000). Federal Ministry for Development, science and the environment, Belgrade

Gallup Balkan Monitor Insights and Perceptions: Voices of the Balkans – 2009 Summary of Findings. In: Partnership with the European Fund for the Balkans. http://www.balkan-monitor.eu/files/BalkanMonitor2009_Summary_of_Findings.pdf. Accessed 20 Nov 2009

Gleni M (2001). Balkan: 1804–1999: nacionalizam, rat i velike sile. (The Balkans 1804–1999. Nationalism, War and Great Powers). Beograd: B92

ISO (2009) ISO Survey 2007. http://www.iso.org/iso/survey2007.pdf. Accessed 28 Nov 2009

Koyama Y (2003) South Eastern Europe in transition, a quest for stabilization of the region after the breakup of the former Yugoslavia. Niigata University, Japan

Myers N (2009) Environmental security concerns: sources. In: Stec S, Baraj B (eds) Energy and environmental challenges to security. Proceedings of the NATO advanced research workshop on energy and environmental challenges to security, Budapest, 21–23 November 2007 (pp 41–56). AA Dordrecht. Springer, the Netherlands

O'Brien K, Leichenko R (2007) Human security, vulnerability and sustainable adaptation. Occasional paper. Human development report 2007/2008, Fighting climate change: human solidarity in divided world. UNDP, New York

REC (2001) International law and the Baia Mare cyanide spill. Regional Environmental Centre for Central and Eastern Europe, Szentendre

REC (2006) Highlights of the regional environmental reconstruction programme for South Eastern Europe. Regional Environmental Centre for Central and Eastern Europe, Szentendre

SEC (2009a) Albania 2009 progress report, Brussels, 14.10.2009, 1337

SEC (2009b) Bosnia and Herzegovina 2009 progress report, Brussels, 14.10.2009, 1338

SEC (2009c) Croatia 2009 progress report, Brussels, 14.10.2009, 1333

SEC (2009d) Kosovo under UNSCR 1244/99 2009 progress report, Brussels, 14.10.2009, 1340

SEC (2009e) Macedonia FYR – 2009 progress report, Brussels, 14.10.2009, 1335

SEC (2009f) Montenegro 2009 progress report, Brussels, 14.10.2009, 1336

SEC (2009g) Serbia 2009 progress report, Brussels, 14.10.2009, 1339

Snoy B, Baltes M (2009) Environmental security: a key challenge for the OSCE, pp. 314–316. http://www.core-hamburg.de/documents/yearbook/english/07/SnoyBaltes-en.pdf. Accessed 25 Oct 2009

Stewart EJ (2004) Preventing and managing conflict in the former Yugoslavia. In: Fatić A (ed) Security in Southeastern Europe. The Management Center, Belgrade, pp 223–248

Stojković M (1997) Geopolitički i strateški činioci savremenih balkanskih odnosa (Geopolitical and strategic factors of contemporary Balkan relations). In: Stojković M, Damian A (eds) Savremeni procesi i odnosi na Balkanu (Contemporary processes and relations in the Balkans). Institute of International Politics and Economics, Centre of International Studies – Faculty of Political Sciences, Belgrade, pp 149–160

Sugar PF, Lederer IJ (eds) (1994) Nationalism in Eastern Europe. University of Washington Press, Seatle & London

Todić D (1999) Okvirna pitanja odnosa medjunarodne bezbednosti i bezbednosti životne sredine, (General issues relevant for relations between international security and environmental security) *Vojno delo*, Beograd, No. 5-6/99, p 47–63

Todić D (2008) "Ekološki menadžment u uslovima globalizacije" (Environmental management and globalization). Megatrend University, Belgrade, 2008

Todić D, Vukasović V (2009) Regional security and international legal regulation of water protection and use – the case of Serbia. In: Stec S, Baraj B (eds) Energy and environmental challenges to security. Springer, AA Dordrecht, pp 141–156

UNDP (2007a) Human development report 2007/2008, fighting climate change: human solidarity in divided world. UNDP, New York

UNDP (2007b) Environmental policy in South Eastern Europe. UNDP, Belgrade

UNECE (2003) New threats to security in the economic social and environmental dimensions (2003). A UNECE Report, Eleventh OSCE Economic Forum, Prague, 23–25 May 2003, Prague, Czech Republic

UNEP (1999) Kosovo under UNSCR 1244/99 – The Kosovo conflict – consequences for the environment and human settlements. UNEP, Nairobi

UNEP (2002) Serbia and Montenegro – Depleted Uranium in Serbia and Montenegro, Post-Conflict environmental assessment in the Federal Republic of Yugoslavia. UNEP, Nairobi

UNEP (2003a) Bosnia and Herzegovina – depleted uranium in Bosnia and Herzegovina, Post Conflict Environmental Assessment. UNEP, Nairobi

UNEP (2003b) Environment and security-transforming risks into cooperation. The case of Central Asia and South Eastern Europe, UNEP, UNDP, OSCE

UNEP (2004a) Environment and security consultations in South Eastern Europe, Skopje. The Former Yugoslav Republic of Macedonia, 23–24 Sep 2004, UNEP, UNDP, OSCE, NATO Final Report

UNEP (2004b) Serbia and Montenegro – from conflict to sustainable development: assessment and clean-up in Serbia and Montenegro. UNEP, Nairobi

Vukasović V (1997) Ekološka bezbednost – jedno od ključnih pitanja saradnje na Balkanu (Environmental security – one of the key issue of Balkan cooperation). In: Stojković M, Damian A (eds) Savremeni procesi i odnosi na Balkanu (Contemporary processes and relations in the Balkans). Institute of International Politics and Economics, Centre of International Studies – Faculty of Political Sciences, Belgrade, pp 257–276

Vuori L (2001) Environmental security in South-eastern Europe: a basis for regional cooperation. In: Schnabel A (ed) Southeast European security, threats, responses and challenges. Nova Science Publishers, New York, pp 23–40

World Bank (2008) South Eastern Europe disaster risk mitigation and adaptation programme, World Bank, ISDR

Xenakis S (2004) Organized crime in the Balkans: pitfalls of threat assessment. In: Fatić A (ed) Security in Southeastern Europe. The Management Center, Belgrade, pp 197–212

The Impact of International Treaties on Climate Change in SEE Countries

Massimiliano Montini

Abstract The paper is divided in three parts. In the first one, the climate change issue is presented and a brief analysis of the major international treaties on climate change is provided. A special focus is placed on the concept of the combination between mitigation and adaptation activities which is required to deal effectively with climate change.

The second part deals with the link between climate change and environmental security, stressing in particular the contribution that climate change may give to worsen the already occurring environmental crisis at a global level, acting as a "threat multiplier". In order to prevent the effect of such a possible combined negative effect of climate change and environmental security, a serious and sharp series of climate change mitigation and adaptation activities must be performed in the years to come in an even more constant and incisive way compared to the last two decades.

The third part addresses more specifically the issue of how to reduce environmental security risks in SEE through the implementation of climate change treaties. In such a context, in particular, two specific scenarios are proposed and analysed.

Firstly, under option A, the simpler possibility for SEE countries to implement the existing major climate change treaties, namely the UNFCCC and the Kyoto Protocol, is considered. The formal implementation of such treaties should be accompanied by the development of appropriate national mitigation and adaptation policies to effectively tackle climate change and prevent climate change related risks. Depending on the seriousness and effectiveness of such national policies, the outcomes of the domestic actions in each of the SEE countries may finally prove more or less successful.

Secondly, under option B, a more sophisticated integrated approach is proposed for the SEE countries, based on the track indicated by the 2005 *Energy Community South East Europe Treaty* (ECSEE Treaty), which aims *inter alia* at reducing environmental risks by establishing a single and comprehensive regulatory framework for trading energy across SEE. Such a treaty imposes to the contracting parties, as a prerequisite for its effective enforcement, the implementation of the most relevant

---

M. Montini (✉)
University of Siena, Department of Economic Law,
Piazza S. Francesco, 7, 53100, Siena, Italy
e-mail: montini@unisi.it

acquis communautaire in the environmental, energy and competition fields. Moreover, it calls for the ratification of the Kyoto Protocol on climate change, thus linking within a broad comprehensive legal framework the major climate change, environmental, energy and competition issues in an integrated way.

Keywords Climate change • Multilateral environmental agreements (MEAs) • Implementation • UNFCCC • Kyoto Protocol • Mitigation and adaptation • Environmental security • Energy Community Treaty

1 Climate Change: Challenges, Responses, Risks and Opportunities

1.1 The Climate Change Issue and the Institutional Response

The climate change issue arose in the late eighties, once the available scientific evidence started suggesting that the progressive increase in the greenhouse gases concentrations in the atmosphere may contribute to a large extent to the "greenhouse effect." The greenhouse effect is a natural phenomenon, caused by a range of different gases among which water vapour and CO_2. It determines the heat emitted from the earth's surface to be retained within the earth's atmosphere.[1]

The progressive rise in the concentration of CO_2 in the atmosphere since the industrialization era[2] increased the natural greenhouse effect and, as a consequence, the earth is becoming warmer.[3] This gives rise to the so-called climate change phenomenon.

According to the findings of the IPCC Fourth Assessment Report (2007), climate change is having discernible physical impacts on physical and biological ecosystems, including natural systems related to snow, ice and permafrost; terrestrial biologoical systems (fauna and flora); marine and freshwater biological systems; agriculture and forestry; human health.[4]

In more general terms, it can be argued that climate change negatively affects ecosystems and makes it increasingly difficult to manage natural resources in a sustainable way, taking into account the needs of the present without compromising the ability of future generations to meet they own needs.

---

[1] The earth's average temperature is at the moment around 14°C. If the natural greenhouse effect did not exist, the average temperature would be around minus 19°C.

[2] At present, the concentration of CO_2 in the atmosphere is about 385 ppm. Before the industrialisation period it was about 280 ppm.

[3] For instance, according to the 2007 IPCC Report, the earth's average temperature has risen by 0.74 degrees in the period from 1906 to 2005. The warming is stronger over land areas than over the sea, and accordingly it is strongest in the north. At the same time occurrences of heat waves and violent downpours have also increased, the oceans have risen, and the ice at the world's poles and on its mountains has begun to melt.

[4] See 2007 IPCC Report, Synthesis Report, p. 7 ff.

The institutional response to the climate change issue at international level has been quite rapid and efficient over the last 20 years, although it may be questioned whether it has been sufficient to effectively tackle the climate change challenge.[5] The first reaction of the international community to the climate change issue dates back to the eighties of last century, when the general Assembly of the United Nations adopted the Resolution 43/53 of 1998 which defined climate change as a "common concern of mankind" to be addressed by the international community with a coordinated action at a global level.[6]

In the same year, the World meteorological organization (WMO) and the United Nations environment programme (UNEP), established the Intergovernmental Panel on Climate Change (IPCC), with the task of analyzing and verifying in an objective, open and interdisciplinary way the state of scientific analysis related to climate change, in order to provide reliable evidence upon which the international community could base its institutional and legal efforts to efficiently tackle climate change.

The IPCC, since then, has correctly and effectively performed its task of main "evaluator" of the existing climate change evidence, which had been conferred to it by the international community, by involving most of the more prominent experts in this field at international level. In particular, the IPCC has produced so far four general reports on the state of science of climate change, issued respectively in 1990, 1995, 2001 and 2007.[7] Such reports are not the results of the own scientific research performed by the IPCC, but they rather represent the outcome of the independent evaluation performed by the IPCC experts on the existing interdisciplinary scientific evidence on the matter.

1.2 The UN Framework Convention on Climate Change

The first IPCC Report, published in 1990, confirmed that a rise in the temperature was already occurring and warned about the possible risks associated to this phenomenon. The IPCC findings, in fact, paved the way for a rapid negotiation of the first international legal instrument related to climate change, then started with the UN General Assembly Resolution 45/212 of 1990 which created an intergovernmental negotiating committee with the aim of drafting an international convention to fight against climate change.[8]

Two years later, the result of the work of the negotiating committee was presented to the UN conference on environment and development, which took place in

---

[5] On the climate change regime in general see Douma-L and Montini (2007); Peeters and Deketelaere (2006); Metz and Hulme (2005); Verheyen (2005); Freestone and Streck (2005); Bothe and Rehbinde (2005); Yamin and Depledge (2004); Victor (2004).
[6] See Resolution United Nations General Assembly (UNGA) 43/53 (1988).
[7] See the IPCC reports on the web site http://www.ipcc.ch.
[8] See Resolution United Nations General Assembly (UNGA) 43/212 (1990).

Rio in June 1992. During the conference the text of the UN Framework Convention on climate change (UNFCCC) was officially concluded and opened to signature.[9]

The international community as a whole gave a clear sign of its willingness to effectively tackle the climate change challenge through the rapid signature of the convention by over 80 countries in a few months. The Convention officially entered into force in 1994 and now it has 194 Parties, including the European Union together with its Member States.[10] This very broad membership makes the Convention a real instrument of global application with a very high level of participation of different stakeholders such as States, NGOs and other international organizations.

The preamble of the UNFCCC, recalling the already mentioned UN Assembly General Resolution 42/212, defines climate change as a common concern of humankind and, taking into account the global nature of the climate change phenomenon, calls all the parties to the widest possible cooperation for a common objective, each one on the basis of the respective possibilities, capabilities and economic and social conditions, pursuant to the principle of the common but differentiated responsibilities, which is embodied in the Convention.

The main objective of the UNFCCC is contained in article 2 and refers to the stabilization of GHGs concentration to the 1990 level, or more properly to the *"stabilization of greenhouse gas concentrations in the atmosphere at a level that would prevent dangerous anthropogenic interference with the climate system"*. Such a level, according to the provisions of the UNFCCC, *"should be achieved within a time frame sufficient to allow ecosystems to adapt naturally to climate change, to ensure that food production is not threatened and to enable economic development to proceed in a sustainable manner."*

The most relevant environmental legal principles upon which the parties should base their action when implementing the Convention, as enshrined in article 3 thereof, are the principle of intergenerational equity, the principle of common but differentiated responsibilities, the precautionary principle and the principle of sustainable development.[11]

In particular, the principle of intergenerational equity requires the Parties to fight against climate change for the benefit of the present and future generations, whereas the common but differentiated responsibilities implies that all Parties have to give their contribution to meet the objectives of the Convention, but their specific obligations should be defined in accordance to their respective historical responsibilities in the increase of GHG concentration in the atmosphere and their present economic, financial and technological capabilities to tackle climate change issues. The principle of common but differentiated responsibilities is supplemented by the principle of equity, which requires the Parties to take into account the specific needs and circumstances of the developing countries in the definition of the respective obligations and duties.

---

[9] See UN Framework Convention on climate change (UNFCCC), New York (1992), in International Legal Materials, vol. 31 (1992), p. 849 or in the web site http://www.unfccc.int.

[10] Updated 1 January 2010 (see http://www.unfccc.int).

[11] On the most relevant environmental legal principles see Montini (2008); Yamin and Depledge (2004), cit., p. 66.

The Convention is also deeply rooted in the precautionary principle, which calls for the adoption of anticipatory measures to prevent or minimize the causes of climate change and mitigate its adverse effects, even in the absence of full scientific certainty about the exact links between the increase of the anthropogenic GHG emissions and their effects on global warming. In fact, the relevance of such a principle was much higher in the early years of implementation of the Convention while is becoming increasingly lower as the scientific evidence given by the IPCC reports shows the closer interaction between anthropogenic GHG emissions increase and the phenomenon of climate change.

Moreover, in the implementation of the Convention the Parties have a right to, and should, promote sustainable development. This means that the specific actions to tackle climate change should be integrated in the national sustainable development policies of the Parties.

The subsequent article, namely article 4(1), contains the main obligations placed upon the Parties following the general objective of the stabilization of GHG concentrations foreseen by article 2 and taking into account in particular the principle of common but differentiated responsibilities. The international efforts to fight against climate change should be led by the most industrialized countries, including the Parties with economies in transition, as listed in Annex I to the UNFCCC (Annex I Parties). The two main obligations listed in article 4(1) for all Parties to the Convention are the following: (i) develop, periodically update, publish and make available to the Conference of the Parties, in accordance with Article 12, national inventories of anthropogenic emissions by sources and removals by sinks of relevant greenhouse gases, using comparable methodologies to be agreed upon by the Conference of the Parties; (ii) formulate, implement, publish and regularly update national and, where appropriate, regional programmes to mitigate climate change and facilitate adaptation to climate change.[12]

1.3 *The Kyoto Protocol on Climate Change*

The Framework Convention, which pursues the general objective of the stabilisation of GHG concentrations in the atmosphere over the long term and contains the basic principles and rules upon which the climate change legal regime is based, was meant to represent the starting point for future more stringent actions, to be contained in subsequent legal instruments.

Shortly after the entry into force of the Framework Convention, negotiations started for drafting a Protocol containing more precise GHGs reduction commitments for the most industrialised countries, namely those already listed in Annex I to the Convention. After two years of intense negotiations, the Kyoto Protocol was finally concluded by the UNFCCC Parties in the framework of COP-3 held in Kyoto in 1997.

---

[12] See article 4(1)(a) UNFCCC on inventories and 4(1)(b) on national programmes.

The Protocol foresees for the first time binding reduction commitments of GHG emissions for the most industrialized Parties (Annex I Parties) and represents the first concrete step for the achievement of the ultimate objective envisaged by the Framework Convention, namely the stabilization of GHG emissions in the atmosphere at a level which that would prevent dangerous anthropogenic interference with the climate system in accordance with the provisions of article 4(2)(a) and 4(2)(b) of the Framework Convention. As to the developing Parties, namely the Non-Annex I Parties, the Kyoto Protocol confirms their full participation to the international efforts to fight against climate change, but does not foresee any specific binding reduction commitment for such Parties.

The reduction commitments established by the Kyoto Protocol for the Annex I Parties are expressed in a reduction percentage as compared to the 1990 levels and are related to a specific deadline, the so-called first commitment period which ranges from 2008 to 2012. The Annex I Parties should reach an overall reduction of 5.2% of their emissions as compared to the 1990 levels. In order to do so, each of the Annex I Parties has a differentiated target as listed in Annex B to the Kyoto Protocol. Not all targets listed there, however, correspond to GHG reduction commitments. In fact, whereas most of the countries have in fact (differentiated) reduction targets, such as for instance, the European Community (then European Union) and their (then) 15 Member States (–8%), the USA (–7%), Canada and Japan (–6%), a few countries have a stabilisation target to the 1990 levels (Russia and Ukraine) and a few others have a limited increase target (Australia +8%).[13]

With regard to the EC (then EU) and its Member States, in particular, article 4(1) of the Kyoto Protocol foresees a peculiar mechanism which enables the "economic integration organisation" to reach an (internal) agreement with its Member States in order to redistribute the overall reduction target –8% among the 15 countries concerned through the determination of specific differentiated targets for each of them. This burden sharing mechanism has been named the "EU bubble". The internal agreement needed to make the mechanism work, the so-called "Burden Sharing Agreement", was in fact concluded in 1998 among the environmental ministries of the then 15 Member States and subsequently notified to the Climate Change Secretariat.[14]

To this respect, it should be noted that, in general terms, the EU bubble aims at promoting a more equitable redistribution of the overall –8% reduction objective, by taking into account the different levels of development and the different capabilities of the 15 EU Member States. In fact, the "Burden Sharing Agreement" foresees mostly reduction targets, which range for instance from the –21% of Germany and Denmark to the –12.5% of the UK to the –6.5% of Italy, alongside

---

[13] See Annex B of the Kyoto Protocol.
[14] The EU Burden Sharing Agreement reached within the European Council of 16–17 June 1998 (EC Council Conclusions DOC 9702/98 of 19 June 1998) was then included in the EC Council Decision 2002/358 of 25 April 2002 concerning the approval, on behalf of the European Community, of the Kyoto Protocol to the United Nations Framework Convention on Climate Change and the joint fulfilment of commitments thereunder (OJ L130, 15 May 2002, p. 1).

stabilisation objective to the 1990 levels (France and Finland) and a limited increase targets for the (then) less developed economies, such as Ireland (+13%), Spain (+15%), Greece (+25%), Portugal (+27%).

As it is well known, the refusal of the USA to ratify the Kyoto Protocol, despite their initial (political) willingness to do so, delayed for several years its entry into force, which occurred just in February 2005. However, during the last few years, the initial gap between the broad UNFCCC membership and the more limited one of the Kyoto Protocol has been constantly reducing and presently the total number of Parties to the Protocol has reached 190, including the European Union together with its Member States.[15]

The specific GHG reduction commitments for Annex I Parties listed in Annex B to the Kyoto Protocol do not exhaust all the obligations of the Parties under the Protocol. In fact, in more general terms, article 10 reaffirms the existing commitments contained in article 4(1) of the Framework Convention and places upon all Parties to the Kyoto Protocol the duty to *"continuing to advance the implementation of these commitments in order to achieve sustainable development"*. In practical terms, the most relevant provisions in this sense is contained in article 10(b)(i) of the Kyoto Protocol which places upon the Parties the duty to *"formulate, implement, publish and regularly update national and, where appropriate, regional programmes containing measures to mitigate climate change and measures to facilitate adequate adaptation to climate change"*.

Moreover, within the process for determining such appropriate national policies for mitigation and adaptation, a special attention has to be placed on the possible impacts of such measures on the traditional national policies in several key national sectors, such as energy, transport, industry, agriculture, forestry and waste management.

1.4 The Mitigation and Adaptation Policies

The main focus of the activities to be planned and performed to effectively tackle the climate change issue therefore consists in the development and enforcement of appropriate national policies and measures for mitigation and adaptation, although it should be recalled here that the Parties with binding reduction committments may also make use of the so-called flexibility mechanisms, foreseen by the Kyoto Protocol, namely Joint Implementation, Clean Development Mechanism and Emission Trading, to partially contribute to the fulfillment of their obligation commitments.[16]

The focus of the present analysis however consists primarily in the national policies and measures that the Parties have to develop at national level, in order to

---

[15] Updated 1 January 2010 (see http://www.unfccc.int).

[16] On the opportunities arising from the use of the flexibility mechanisms and in particular from the Clean Development Mechanism (CDM), with a particular focus on the Western Balkan countries, see Montini 2010.

contribute to the global effort for mitigation and adaptation to climate change. In this sense, it seems that a couple of basic questions need to be firstly answered. They read as follows: which kind of risks and opportunities exist with regard to the development and enforcement of appropriate national climate change policies and measures for mitigation and adaptation? How different ought to be the approach of developed and developing country Parties with respect to these two kinds of measures?

As to the first question, the development and implementation of climate change national policies should not be seen simply as an instrument to serve the purpose of meeting the Kyoto Protocol obligations and contribute to the global effort against climate change, but they should be rather considered as an opportunity, rather than a burden, placed on the parties, which may provide a terrific incentive to restructure their economies with a low-carbon scenario in mind, thus contributing to improve the efficiency as well as the sustainability of their industrial and energy systems, attract foreign direct investments and enhance the competitiveness of their national economic systems.

Moreover, despite the fact that putting in place adequate mitigation and adaptation policies may be costly for all Parties, it has be clearly shown, most notably by the *Stern Review*,[17] that the deployment of an early action against climate change may actually prove a reasonable choice also in purely budgetary terms, insofar the cost of inaction or late action in this field may be finally prove much more costly in the long term.

As to the second question, with reference to the mitigation policies, although the Kyoto Protocol imposes binding reduction committments only upon the most industrialised countries (Annex I Parties), and despite the fact that even in the near future the acceptance of binding emission reduction limits for most Non Annex-I Parties countries may be neither desirable, nor feasible, on the basis of what we have been saying above, it should be clear that for all countries, irrespective of their Annex-I or Non-Annex I membership, some restructuring of their national economies with a low-carbon scenario in mind, may be a successful choice in a medium-long term perspective, insofar it may increase their competitiveness and improve their overall sustainability.

In addition to that, with regard to the adaptation policies, it should be underlined that even for Parties which are neither already engaged, nor are to be involved in the next few years into active mitigation policies, the development and effective enforcement of broad and sharp adaptation policies may be a necessary option to be pursued in order to minimise environmental risks, improve resource management and increase sustainability.

For all the above mentioned reasons, all Parties to the Kyoto Protocol in the next few years should seriously engage in planning and implementing mitigation and adaptation policies, not necessarily being subject to emission reduction limits, nor being bound to the achievement of specific and measurable mitigation targets, but trying to get away from traditional "business as usual" development scenarios, which in many cases are largely based on high energy intensive models. Low-carbon

---

[17] See Stern (2007).

economy patterns of (sustainable) development should be pursued instead and the enforcement of appropriate national mitigation and adaptation policies in all Parties may help going into the right direction.

2 The Link Between Climate Change and Environmental Security

2.1 What Is Environmental Security?

There is not a commonly accepted comprehensive definition of environmental security. With this term, one normally refers to the complex relationship between all the possible alterations of the status of the environmental media and the natural ecosystems on the one side and the risks for the security of individuals and communities on the other side.

When one talks about environmental security, the reference is for instance to all the possible negative effects that the humanity is perpetrating to itself, by mismanaging natural resources as well as by polluting air, water and soil. The analysis of the cause-effect relationship between anthropogenic environmental degradation and natural resources mismanagement and the increase in environmental security risks should include for instance a reference to the growing scarcity of natural resources, which makes them actually much more difficult to be managed in a sustainable way; to the progressive depletion of air, soil and water (sea and freshwater) resources, both in terms of quality and quantity; to the increasing land degradation and desertification; to the rising difficulty to have an adequate access to food, health and the basic living conditions for many people around the world.

In such a context, moreover, the specific factors which negatively affect environmental security risks may lead in some cases to situations of "destabilisation," both at local or national as well as at international or global level. To this respect, it should be mentioned that, although so far no clear evidence of environmentally related "armed conflicts" has been detected, it cannot be a priori excluded that in the future some environmental related "stresses" may even give rise to armed conflicts in same areas of the world.

2.2 How Climate Change Can Act as "Threat Multiplier"

Climate change certainly represents the main challenge in terms of environmental security for various reasons.

Firstly, since it has a genuine "global" dimension and it imposes a shift from a State or regional security perspective to a "global security perspective," which regards all humanity, although its negative consequences may be perceived to be more severe in some countries than in others.

Secondly, since it may give a particularly negative contribution in terms of worsening the already occurring global environmental crisis (e.g. environmental degradation and reduction in quality and quantity of natural resources), insofar it may cause sharp modifications to the basic environmental and living conditions in many areas of the planet, thus reducing the capacity of individuals and communities to adapt rapidly and effectively to the increasing threats. In such a context, climate change may in fact act as a "threat multiplier," that is as a factor which is potentially able to exacerbate existing sources of conflict and insecurity which are somehow related to the progressive environmental pollution and depletion of natural resources.

In such a sense, climate change may act as a "threat multiplier," with respect to environmental security, through several channels. The 2009 Report of the UN Secretary General on *"Climate change and its possible security implications,"*[18] which was prepared in response to the UN General Assembly Resolution 63/281,[19] and is based on the views of several Member States and relevant regional and international organisations, quite interestingly identifies five channels through which such a process may effectively take place:

(a) The first one is named "vulnerability". Through this channel the impacts which may be caused or increased by climate change to the human well-being are considered. In this context, the focus is in particular on the most vulnerable communities which are threatened in terms of reduced food production due to the adverse conditions that climate change will most probably create, by reducing for instance cereals productivity in some parts of the world. Food security will become a serious issue for many poor people in developing countries and this may give rise in turn to social protests and unrest around the world. Similarly, shortage of water will become a severe question at a global level, with potentially high repercussions on human well-being, human health, human habitability of certain islands and hydro-power production. Moreover, sea level rise associated with climate change could make some coastal areas uninhabitable, causing loss of habitat and livelihood for many people and increasing the risk of massive displacement of populations due climate change reasons.

(b) The second channel is called "development." Under this item, all the possible negative impacts that climate change may have in terms of slowing down the economic development process in several countries are considered. This may occur for instance through crop losses due to more frequent extreme weather events, such as droughts and floods, which may reduce economic output, beside reducing human well-being. This may in turn cause negative effects on nutrition and human health, thus affecting human productivity. Moreover, it may cause land degradation and coastal erosion, provoking damages to man-made infrastructures with sensible economic consequences on certain countries' productivity. These effects may be particularly severe in developing countries which are more heavily dependent on primary sectors, such as

---

[18] See United Nations, Report of the Secretary General, A/64/350, 11 September 2009.
[19] See UN General Assembly Resolution A/RES/63/281, 11 June 2009.

agriculture, fisheries and forestry, and may lead to an increased poverty and the reversal of some development gains in many countries, which may slow down the process towards the achievements of the Millennium Development Goals.
(c) The third channel refers to "coping and security" issues. In particular, under this item the major concern is placed on the possible security risks that may be associated with uncoordinated coping and survival strategies devised by populations severely affected by climate change related disruptions to their ecosystems, societies and economies. In such a context, in fact climate change impacts may induce voluntary or involuntary migration or in some cases also give rise to competition over scarce resources, possibly giving rise also to armed conflicts, with both domestic and international dimensions.
(d) The fourth channel relates to "statelessness" risks. This refers to the implications related to the very ultimate threat for some countries, which is the disappearance of their territory due to sea-level rise. This is a concrete and serious issue for many small island developing States around the world, such as for instance the Maldives, where 80% of the land is less than meter above sea level. In case of loss of their territory, it is to be assumed that certain States may disappear, both in physical and in legal terms, with the risk of mass statelessness of their populations. In this sense, some multilateral agreements should be drafted as a preventive mechanism, so as to ensure an adequate and fair treatment to populations in case of disappearance of their territories.
(e) The fifth channel is named "international conflict" and relates to the potential negative effects that climate change may cause to international relations, in particular by possibly exacerbating the risks of conflicts over the use of trans-boundary resources, such as international rivers, as well as over global resources previously not accessible, such as the Artic sea natural resources deposits and maritime transport routes, which may become available due to climate change related effects and events. However, with respect to this issue, the Report of the UN Secretary General shows a certain optimism that the challenges posed by climate change will rather reinforce international relations and increase cooperation over intensified scarce resources.

2.3 *How to Prevent the Combined Negative Effect of Climate Change and Environmental Security?*

In order to prevent climate change to act as a "threat multiplier" some specific "threat minimizers" can be identified, namely in the form of specific actions which beside being per se right patterns to be pursued, in the present context may serve the most relevant scope of reducing the environmental risks associated with climate change. To this effect, the already mentioned Report of the UN Secretary General on "*Climate change and its possible security implications*" identifies five "threat minimizers", which may help in reducing the climate change related environmental security risks.

They consist in the promotion of the following specific actions:

(a) Effective international and national mitigation actions, supported by finance and technology flows from developed to developing countries; such efforts should be supported by a strong multilateral legal regime, based on the UNFCCC and the (revised) Kyoto Protocol or its successor. Mitigation activities should be coupled with a move towards low-carbon economy models, both in developed and in developing countries. However, in order to achieve this objective, the deployment of an adequate financial assistance as well as the effective flow of new and cleaner technology will be essential, in particular to support less developed countries.

(b) Strong support to adaptation and related capacity-building in developing countries; in this sense, adaptation refers to empowering people to react to climate change threats, helping them to improving their resilience, securing their livelihoods and strengthening their physical infrastructures to protect against extreme events. Moreover, it consists in promoting capacity-building to increase the strength of their institutions, as well as their legal and political systems, to prevent major risks in terms of food security, health and safety.

(c) Inclusive economic growth and sustainable development, which will be critical to building resilience and adaptive capacity; based on the assumption that a continuous economic growth is essential to strengthen resilience and promote adaptation to climate change, as well as to foster social cohesion and limit the risks of social conflicts. States should try to encourage "climate-proofing" economic development, taking into account also the need to promote more sustainable patterns of development and to ensure that, in particular for less developed countries, climate change considerations are fully integrated in the national plans to achieve the Millennium Development Goals.

(d) Effective governance mechanisms and institutions; in this context, it is stressed that the existence, at local or national as well as at regional or international level, of mechanisms and institutions to promote an effective governance of the climate change relate threats is essential to respond to the emerging security risks. National governments should be empowered to develop and implement effective and coordinated climate change national strategies, whereas all sectors of society should be engaged in such efforts.

(e) Timely information for decision-making and risk management. So far, the findings of the IPCC Reports have represented a crucial basis to improve the understanding of the climate change phenomenon, foresee the associated risks and develop a sound and informed decision-making process on mitigation and adaptation measures. In the future, it will be essential to further enhance earth observation and climate monitoring systems, as well as to make timely and adequately available the relevant information, so as to support an effective decision making at all levels.

In addition to the description of the above mentioned five patterns of actions, which may act as "threat minimizers," the Report of the UN Secretary General also stresses that international cooperation should be reinforced to properly address transboundary effects and to prevent and resolve climate-related conflicts in accordance with the Charter of the United Nations.

Looking from a general perspective to the list of the "threat minimizers" proposed by the UN Secretary General, it emerges clearly the paramount role which is assigned to the development of national climate change policies and the performance of a series of mitigation and adaptation actions in all countries. In particular, it seems that in order to prevent climate change from increasing the existing sources of environmental security risks, a serious and sharp series of climate change mitigation and adaptation activities should be performed in the years to come by all countries around the world in a more consistent and incisive way as compared to the last two decades.

For instance, it seems that, on the one side, in terms of mitigation, even those countries which do not have binding emission limitation and reduction committments under the Kyoto Protocol should try to pursue domestic policies aimed at progressively reducing their overall emissions, and in particular their per capita emissions and energy intensity. On the other side, with regard to adaptation, all countries should try to develop national plans containing specific measures aimed at preventing the most severe possible risks of damage to their environment and their natural resources. In fact, in both cases, this might have for them a positive effect both in economic and budgetary terms as well as in environmental and sustainability terms.

3 How to Reduce Environmental Security Risks in SEE Through the Implementation of Climate Change Treaties?

3.1 Implementing Climate Change Treaties in SEE

Generally speaking, SEE countries do not give a great contribution to climate change. In fact, their overall production of GHG emissions is minimal in absolute terms.[20] However, they mostly have outdated industrial and energy production systems and their energy grids are often inadequate to cope with an increased and more diffuse production of electricity, which may be driven for instance by the possible expanded recourse to some renewable energy production sources (e.g. photovoltaic and wind energy sources).

Therefore, they should consider climate change treaties obligations as an opportunity to revise their outdated systems, rather than a burden imposed on their economies. In fact, climate change related investments may bring a modernisation to their national production and energy systems.

Moreover, the pursuit of a sharp action in mitigation and adaptation activities to tackle climate change may concretely help such countries to reach also other very relevant objectives.

---

[20] Between 1990 and 2003, GHG emissions in SEE increased by about 2% after a strong decrease during the first half of the 1990s. Emissions from SEE in 2004 were 599 Mtonnes (source: Europe's Environment – the Fourth Assessment by European Environmental Agency, Copenhagen, 2007).

Firstly, it may help making their economies better equipped to respond to the low-carbon economy challenge, which is becoming the most relevant question for all countries around the world both in economic, energy, social and environmental terms.

Secondly, it may help reducing environmental security risks, in the terms highlighted above, provided that SEE countries manage to attract enough financial support, adequate foreign direct investments and an increased transfer of cleaner technologies to implement their national climate change policies and promote the shift to a more sustainable economy.

Therefore, summing up, it can be argued that environmental security risks in SEE countries may be greatly reduced through a correct and pro-active implementation of the most relevant climate change treaties. In such a context, in particular, two specific scenarios are proposed and analysed.

Firstly, under option A, the simpler possibility for SEE countries to implement the existing major climate change treaties, namely the UNFCCC and the Kyoto Protocol, is considered.

Secondly, under option B, a more sophisticated integrated approach is proposed for SEE countries, which is based on the track indicated by the 2005 *Energy Community South East Europe Treaty* (ECSEE Treaty).

3.2 Scenario A: UNFCCC and Kyoto Protocol Implementation

Under the scenario A, the simpler possibility for SEE countries to implement the existing major climate change treaties, namely the UNFCCC and the Kyoto Protocol, is considered. As already mentioned above, the formal implementation of such treaties is not enough, but should be accompanied by the development of appropriate national mitigation and adaptation policies to effectively tackle climate change and minimise climate change related risks.

Depending on the seriousness and effectiveness of such national policies, the national action in each of the given countries may finally prove more or less successful. However, in developing and implementing their national policies, SEE countries should consider that most of the mitigation and adaptation actions would not only contribute to the global climate change effort, but would rather entail positive effects also from a local and national perspective, by increasing the overall sustainability and competitiveness of their national economies.

To this effect, the main problems which may limit the effective implementation of the climate change treaties in the said countries, and which should be quickly addressed and solved, may be grouped in the following four categories:

1. Lack of political will (the scientific evidence on the risks associated with climate change, even if supported by a growing public opinion awareness, may not be a sufficient driver for a sharp action if there is not a clear political will to develop and implement national climate change policies and activities).

2. Lack of financial resources (the provision of an adequate financial support is strictly necessary to effectively pursue national mitigation and adaptation policies; new funds should be made available, in particular to Non-Annex I countries, to contribute to the climate change global effort; the conditions for an increased flow of foreign direct investments to sustain climate change related projects and activities should be promoted at international level).
3. Lack of technology (the implementation of meaningful mitigation and adaptation actions greatly depends from the effective transfer of adequate technologies, which may prepare, promote and sustain the progressive shift to a low-carbon economy).
4. Lack of capacity building for public officials and professionals (awareness raising is not sufficient to cope with the complex challenges deriving from the climate change treaties; in many cases, the provision of specific ad hoc training courses and actvities is absolutely necessary for the human resources involved in the implementation and enforcement of climate change agreements).

Should the above mentioned problems be quickly addressed and solved, the implementation of the climate change treaties in the SEE countries, proposed under the scenario A, may be quite successful and may contribute to the global effort against climate change. Moreover, it may help enhancing their progressive shift to a more sustainable, competitive, low-carbon economy and reducing specific environmental security risks in the area.

However, given the peculiarity of the SEE countries, which are all greatly influenced by their approximation process towards the European Union, it seems that a better choice for them would be to pursue a more integrated track, which beside contributing to the global effort against climate change may simultaneously reduce their distance to the EU, following the path traced by the 2005 *Energy Community South East Europe Treaty* (ECSEE Treaty or simply *Energy Community Treaty*).[21] In practice this would consist in the definition of common patterns of activities with the EU and its Member States, in the early implementation of the some relevant EU standards and rules and in the progressive establishment of a common European market extended to the SEE countries, with a focus not limited to the climate change priority, but expanded to a broader perspective.

3.3 Scenario B: Energy Community Treaty Implementation

As already mentioned above, under the scenario B, a more sophisticated integrated approach is proposed for SEE countries, which is based on the track indicated by the *Energy Community Treaty*, which aims *inter alia* at reducing environmental risks by establishing a single and comprehensive regulatory framework for trading energy

---

[21] See 2005 *Energy Community South East Europe Treaty* (ECSEE Treaty) or simply *Energy Community Treaty*, published in EU OJ L 198/18 of 20 July 2006. See also http://www.energy-community.org.

across SEE. Such a Treaty, in fact, aims at establishing an integrated market in the gas and electricity sectors, with a particular focus to the territory of the EU countries which are neighbours to the SEE countries, namely Austria, Greece, Hungary, Italy and Slovenia, and the SEE countries themselves, namely Albania, Bulgaria, Bosnia and Herzegovina, Croatia, the former Yugoslav Republic of Macedonia, Montenegro, Romania, Serbia, the UN Interim Administration in Kosovo.

The integrated market for the gas and electricity sectors is based on a single regulatory framework, which should promote the following objectives:

1. Attract investments in gas networks, power generation, transmission and distribution networks, so that security of supply is enhanced.
2. Provide rules for internal and external trade.
3. Improve the environmental situation relating to the gas and electricity sectors.
4. Ensure market competition.

The regulatory framework for the gas and electricity sectors draws from the EU relevant standards and rules. Therefore, the Contracting Parties to the *Energy Community Treaty* which are not EU Member States are required, as a prerequisite for the its effective enforcement, to proceed with the implementation of the most relevant *acquis communautaire* in the environmental, energy and competition fields, as described in Title II of the Treaty itself, which refers to a specific series of relevant EU Directives.

In particular, as for the environmental field, the *acquis communautaire* on environment for the purpose of the *Energy Community Treaty* refers specifically to some selected legal acts and provisions, which include the EC Directive 85/337 on EIA (and subsequent modifications), the EC Directive 99/32 on the sulphur content of certain liquid fuels (and its subsequent amending Directive), the EC Directive 2001/80 on the protection of the air from large combustion plants and the specific provision of article 4(2) of the EC Directive 79/409 on the conservation of wild birds. In addition to the implementation of the above listed legal acts, the Parties to the Treaty "shall endeavour" to implement the EC Directive 96/61 on Integrated Pollution Prevention and Control (IPPC).

Moreover, in the framework of the *acquis communautaire* on environment, article 13 of the *Energy Community Treaty* states that "*The Parties recognise the importance of the Kyoto Protocol. Each Contracting party shall endeavour to accede to it.*" In fact, in such a context, the ratification (and implementation) of the Kyoto Protocol is recognised not only in environmental terms, within a comprehensive scenario, which includes in a broad comprehensive legal framework the major climate change, environmental, energy and competition issues in an integrated way.

Therefore, the *Energy Community Treaty*, within a typical sustainability scenario, although no explicit mention to the principle of sustainable development is made in the text of the Treaty, aims at promoting economic development and social stability, ensure environmental protection, foster cooperation among States and ultimately promote peace and stability thus contributing to the reduction of environmental security risks in SEE countries.

In sum, should the SEE countries decide to follow the proposed scenario B, their progressive shift to a more sustainable, competitive, low-carbon economy would be greatly enhanced, if compared to the scenario A, and the pertinent environmental security risks greatly reduced.

However, the importance of the *Energy Community Treaty* goes well beyond its specific scope of application, insofar it represents a very interesting integrated model of cooperation between the EU and the SEE countries, which could be extended also to other sectors. In fact, it may greatly help reducing the existing gaps among the EU Members and non-Members, encouraging a more coordinated action at internal and external level, promoting a greater economic development and ultimately minimising the specific environmental security risks which are present in the area.

References

Bothe M, Rehbinder E (2005) Climate Change Policy. Eleven International Publishing, the Netherlands
Douma WTh, Montini M, Massai L (eds) (2007) The Kyoto protocol and beyond: legal and policy challenges of climate change. T.M.C. Asser Press, the Hague
Freestone D, Streck C (2005) Legal aspects of implementing the Kyoto protocol mechanisms. Oxford University Press, Oxford
Metz B, Hulme M (eds) (2005) Climate policy options post 2012: European strategy, technology and adaptation after Kyoto. Climate Policy (Special Issue), 5(3), 243–391
Montini M (2008) The role of legal principles for environmental management. In: Clini C, Musu I, Gullino ML (eds) Sustainable development and environmental management – experiences and case studies. Springer, Dordrecht
Montini M (ed) (2010) Developing CDM projects in the Western Balkans. Legal and Technical Issues Compared. Springer, Dordrecht
Peeters M, Deketelaere K (eds) (2006) EU climate change policy: the challenge of new regulatory initiatives. Edward Elgar, Cheltenham
Stern N (2007) The economics of climate change: the stern review. Cambridge University Press, Cambridge
Verheyen R (2005) Climate change damage and international law. Martinus Nijhoff Publishers, Leiden
Victor D (2004) The collapse of the Kyoto protocol and struggle to slow global warming. Princeton University Press
Yamin F, Depledge J (2004) The International climate change regime. Cambridge University Press, Cambridge

Documents

IPCC Fourth Assessment Report (2007), http://www.ipcc.ch
Resolution United Nations General Assembly (UNGA) 43/53 (1988)
Resolution United Nations General Assembly (UNGA) 43/212 (1990)
United Nations, Report of the Secretary General, A/64/350, 11 September 2009
UN General Assembly Resolution A/RES/63/281, 11 June 2009
UN Framework Convention on climate change (UNFCCC), New York 1992, in International Legal Materials, vol. 31 (1992), p. 849 or in the web site http://www.unfccc.int
Energy Community South East Europe Treaty (ECSEE Treaty) (more simply Energy Community Treaty), published in EU OJ L 198/18 of 20 July 2006

The Impact of the International Treaties on Water Management in South-Eastern Europe

Slavko Bogdanovic

Abstract This paper deals with the legal regimes established by different international treaties on water management in South Eastern Europe (SEE). Brief comments concerning history help in understanding the actual trends in the development of legislation on water management in the SEE region. More than ever before, national systems of water management legislation are under the influence of global, regional (in terms of UNECE) and EU legal frameworks. Neighbouring/riparian/river/river basin/lake basin relations between the SEE countries have been under the influence of the new water management paradigm developed by the EU, which imposes to the national systems of water management to take into account trans-boundary impacts of national water projects/activities. In that context, the paper contains the author's views on open fields for cooperation between the SEE countries, as well as concise considerations on situations that might lead to some disputes between countries concerning their shared waters in the SEE region.

Keywords Water management • Environment and security • SEE • Trans-boundary waters • Water treaties

1 Introduction

In the second part of twentieth century, international treaties were used as instruments resolving for disputes between states which share large rivers in different parts of the world. Those disputes served as an impetus for half-a century long International Law Association – ILA studies of international water law (customary and conventional) that eventually led to drafting and signing of one, and so far the only, global multilateral water treaty, which deals with the law on the non-navigational uses of international watercourses.

S. Bogdanovic (✉)
University Business Academy, Novi Sad, Serbia
e-mail: slavkob@open.telekom.rs

In a divided Europe, before the fall of the Berlin Wall, a process of developing multilateral legal regimes, which have been established by several multilateral conventions applicable to environmental protection (including trans-boundary waters) in the region of Europe, was initiated under the aegis of the Conference on Security and Co-operation in Europe – CSCE and the UN Economic Commission for Europe – UNECE. Those treaties are part of the *acquis communautaire*.

International water treaties were used as a tool for regulating (and in that way avoiding or diminishing possibility for evolving disputes) bilateral/trilateral/riparian/ neighbouring relations concerning their shared waters (international or trans-boundary) between countries in South Eastern Europe (SEE).

Today, those broader legislative requirements are enhanced by the more demanding requirements of the European Union. These requirements encompass a change in the existing/old national systems of legal norms regulating water management in SEE countries and the development of new ones, in accord with the newly defined, composite and advanced EU water management paradigm (comprising both national and trans-boundary aspects). Whereas all the countries in SEE are integrated into or committed to integrate them into the EU, they must transpose, implement and enforce EU legislation (i.e. the entire *acquis communautaire*). Changes in SEE countries' societies occurred in a changed Europe after the fall of the Berlin Wall and the Balkans' wars during 1990s have been driving forces for changes of national legal frameworks defining water management systems. Those changes of national law patterns are strongly (more than ever before) influenced by broader, in some case global, legal requirements.

This paper is the result of an attempt to describe the impact of the international treaties on water management in the SEE countries and to identify the open fields for cooperation, particularly in terms of the establishment of adequate legal regimes by international treaties. Finally, the paper contains a brief indication, on certain situations which, if further developed, might lead to tensions and disputes over waters between SEE countries.

2 Some Notes on History

During 1950s and 1960s, a significant number of water treaties were concluded between the Balkan countries (themselves) and between them and their neighbouring countries.[1] This was particularly specific for the socialist Yugoslavia, which tried to establish a long-term co-operation with its neighbours with regard to different issues related to trans-boundary waters. Such policy was in accordance with the international position of Yugoslavia as a non-aligned country in a divided world, committed to develop and implement principles of peaceful co-existence and friendship among peoples and provide conditions for development and benefit to the entire region.[2]

---

[1] The parties of which were the Macedonia, Albania, Austria, Bulgaria, Greece, Hungary, Romania.
[2] UNEP/GRID-Arendal (2007), p. 46.

The Impact of the International Treaties on Water Management   79

Those treaties, and several additional ones, signed during the 1970s and 1980s, mostly bilateral, but sometimes trilateral and even multilateral,[3] had as their subject matter the establishment of co-operation between national authorities competent for water management, with the aim of dealing with the issues related to river basins or focused on certain rivers. For example, such issues were related to:

- Research in the Vardar River
- Floods by trans-boundary watercourses (Danube, Drava, Tisa, Tamis, Nera and others)
- Drainage and improvement (particularly in bordering areas of Macedonia, Albania, Greece and Hungary)
- Construction of dams and exploitation of hydro-electric power production facilities (on the Drava and Danube Rivers)
- Navigation (on the Danube, Tisa, Drava and Bojana Rivers)
- Fishery in boundary waters (e.g. on the Danube, Tisa and Bojana Rivers and the Skadar/Shkoder, Dojran and Prespa Lakes)[4]

Protection against pollution of waters of a river or a river basin was also a subject matter of international treaties,[5] in certain cases with a specific purpose.[6] Sometimes, bilateral co-operation was designed in a bilateral treaty with the aim of realizing certain hydraulic structure on an international river, supported by international funding, with a prospect of opening significant public works.[7] In the case of the catchment areas of the Soca, Idrijca and Timava Rivers, bilateral relations between Yugoslavia and Italy were regulated by a treaty relating to improvement of economic co-operation between two countries.[8]

Often those treaties contained the legal basis for the establishment of mixed commissions for their implementation. Rare were the cases without provisions on institutional arrangements. What were results of work of those commissions is a separate issue deserving more investigative efforts. Here it could only be indicated

---

[3] For more details see, Bogdanovic (2005), p. 81.

[4] Bogdanovic (2005) e.g. p. 81; UNEP/GRID-Arendal (2007) *op. cit.*

[5] In case of the Tisa River, where, it should not be overlooked, the multilateral treaty (the Agreement on the River Tisa and its Tributaries' Protection Against Pollution, Szeged, Hungary, 1985) was never implemented.

[6] E.g. of implementation of preventive measures aimed at protection of fish and other aquatic life from pollution of waters (in the Danube River). Ensuring of normal migration of migratory fish species in the Danube waters in case of construction of hydro-technical structures, particularly dams, was one of the obligations accepted by the parties of a bilateral treaty, but never respected. Migration of Danube sturgeons and obstacles (Iron Gate dams) on their migratory way pays nowadays more attention through the e.g. engagement of the Council of Europe and its action plan for conservation of these ancient Danube species; Council of Europe (2006) Bourne (2001).

[7] Such is the case of the Canal Vardar/Axios, a joint project between Yugoslavia and Greece, which also was never implemented).

[8] Agreement on Improvement of Economic Cooperation between SFR Yugoslavia and the Republic of Italy, signed on 10 November 1975 at Osimo (Ancona), Italy.

that some of them were (and still are) successful.[9] However, a particular case with an unsuccessful end was the work of the Mixed Commission formed by Yugoslavia and Albania.[10] This Commission terminated its activities in 1986 when dealing with the issue of the breach of the interstate agreement by Albania during construction of the hydro-electric power station *"Drita e Partize"* nearby the town of Fierza.[11]

Besides, former Yugoslavia, as a federal state with a considerable number of watercourses crossing or making the borders between its federal units (republics and autonomous provinces) had developed legal frameworks for signing water compacts between the federal units. Among other cases, two such compacts, concerning the conditions for regulation and use of waters of the catchment area of the Trebizat River, involving the Socialist Republics of Bosnia and Herzegovina and Croatia, and several of their municipalities were signed in 1982.[12] Again, it is a separate issue why this conceptually excellent example of intra-state trans-boundary co-operation was not fully implemented.[13]

No doubt that behind such policy of Yugoslavia and the neighbouring countries, along with a clear awareness that joint activities directed towards management of shared waters could be only beneficial for them, there was a firm commitment to avoid such difficult water disputes as arisen in the period after 1945 over the waters of the Indus River (between India and Pakistan), the Nile River (between Egypt and Sudan), the Jordan River (between Israel and its neighbours) or the Columbia River (between Canada and United States).[14] Those disputes ultimately were resolved through the signature of water treaties (except in case of the Jordan River) with the active assistance of the World Bank, sometimes in the form of a kind of "shuttle – diplomacy".

It is worth noting that those serious water disputes initiated unprecedented work of the International Law Association (ILA) on identifying law (customary, conventional, municipal) applicable on multifaceted subject of shared waters, lasting with short breaks half a century (1954–2004). The work of ILA resulted in the precise definition of a series of reliable legal rules relevant for regulation of interstate

---

[9] E.g.: the Danube Commission, established in the framework of the Belgrade Convention on the Danube Navigation Regime (1948); the bilateral commission competent for the Iron Gate hyrdraulic power production and navigable system (Serbia and Romania); the mixed water commissions formed between Yugoslavia and Hungary (established by the Agreement on Water Economy Issues, signed at Belgrade, 1955) and Yugoslavia and Romania (established by the Agreement on Hydraulic Systems and Boundary Watercourses or those Intersected by the State Borders, signed at Bucharest, 1955).

[10] Established by the Agreement between the Government of the Federal Peoples' Republic of Yugoslavia and the Government of the Peoples' Republic of Albania on Water Economy Issues signed on 5 December 1956 at Belgrade.

[11] Source of information in file with the author.

[12] "Peoples Gazette of the SR Croatia", No. 16/82.

[13] The parties to those compacts agreed that their interest is to harmonize mutually their needs concerning regulation and use of watershed area of the Trebizat River, especially in construction of flood protection systems, protection of water, water supply, provision of water balances and development hydro-electric power facilities.

[14] Agreement for the Full Utilization of the Nile Waters, signed on 8 November 1959 (The Nile Waters Treaty); The Indus Waters Treaty, signed on 19 September 1960; The Columbia River Treaty, signed 17 January 1961.

relations in regards of international waters/drainage basins. Here it should be mentioned that in the famous Helsinki Rules (1966) the principle of reasonable and equitable utilization of waters of an international drainage basin was identified as the primary rule governing international water resources.[15]

The work of the ILA served later as precious source of excellent knowledge in the work of the UN International Law Commission (ILC) on the progressive development of international law on non-navigational uses of international watercourses,[16] which, after 27 years, led to adoption of the UN Convention on the Law of Non-Navigational Uses of International Watercourses. This Convention is the first international legal instrument dealing, on the global level, with the law on non-navigational uses of international watercourses. It was adopted with the UN in 1997.[17] The Convention is not yet in force, lacking 17 more ratifications out of 35 needed. Many international organizations and fora urge countries to ratify the Convention.[18] Once in force, it will significantly contribute to shade a light on the existence of certain rules or principles of customary water law and, in that sense, to foster a better cooperation between states concerning trans-boundary water issues on the global level.[19]

3 The Fall of the Berlin Wall and the Re-composition of the Western Balkans

The fall of the Berlin Wall symbolizes the end of a divided Europe and the end of a bi-polar world. Activities of the UNECE that initiated in 1975 were based on the global policy requirements of the UN and Europe. Particularly, the Helsinki Final Act of CSCE (later OSCE) confirmed the existence in international law of a duty to prevent activities on-going in the territory of one member state to be cause of environmental degradation in the territory of another member state.[20] This led UNECE to the development of several multilateral environmental conventions

---

[15] Charles Bourne (2001), p. 5.

[16] For details on the work of ILC, see Sir Artur watts (1999).

[17] See, the Annex to Press Release GA/9248 of 21 May 1997, dated 22 May 1997. The following SEE countries voted in favour of this Convention: Albania, Croatia, Greece, Romania and Slovenia. Turkey voted against. Bulgaria abstained. Macedonia was absent, whereas BiH and Yugoslavia were not listed in the Annex.

[18] For example, WWF issued a call on governments to bring into force the Convention, which is seen as an adequate legal response tool to the problem of integrated river basin management. http://www.panda.org/what_we_do/how_we_work/policy/coventions/water_conventions/un_watercourses_convention/.

[19] Johan G. Lammers (2008), pp. 2 and 5.

[20] Birnie and Boyle (2002), p. 63.

that regulate a number of key environmental issues. Almost all of them pertain to trans-boundary waters, in one or another way.[21]

As the result of the 1990s wars and the disintegration of Yugoslavia, six new countries in the Balkan region have arisen, i.e. Bosnia and Herzegovina (BiH), Croatia, Macedonia, Montenegro, Slovenia and Serbia. Besides, Kosovo (once the autonomous province in the Socialist Republic of Serbia and the member of the socialist Yugoslav Federation) under the factual protectorate of international community based on the Security Council Resolution 1244, declared its independency in 2008. From 1999, Kosovo is out of the sovereign control of Serbia.

In the SEE region, before the beginning of the disintegration of Yugoslavia, there were at least ten international river basins (excluding the Danube River and its several tributaries). A number of earlier national water resources in the territory of Yugoslavia, have become international, after its disintegration,[22] generating the necessity for the development of international legal regimes specifically applicable to them.[23]

The development of shared water resources, their use for different purposes, their protection and the preservation of their aquatic eco-systems, or protection against their detrimental effects, can be provided in a sustainable way only on the basis of ratified water treaties, which can carefully balance interests and regulate relations between interested states. However, nowadays, in changed conditions in comparison to the times of 1950s or 1960s, a number of world wide accepted policy and soft-law instruments (such as the Stockholm Declaration – 1972, the Report of the Brundtland Commission – 1987, the World Charter for Nature – 1982, the Dublin Principles and Rio Declaration – 1992) as well as some other global legal instruments (e.g. the Convention on Wetlands of International Importance Especially as Waterfowl Habitat, Ramsar 1971), together with the above mentioned UNECE international instruments, jointly form the general policy and legal reference framework. In such a context, the development of new legal regimes, established between new (and old) states and over (old and) new international water resources is on-going.

---

[21] Those conventions are Convention on Environmental Impact Assessment in a Transboundary Context (EIA Convention, Espoo, 1991), Convention on the Transboundary Effects of Industrial Accidents (IA Convention, Helsinki, 1992), Convention on the Protection and Use of Transboundary Watercourses and International Lakes (Water Convention, Helsinki, 1992) and Convention on Access to Information, Public Participation in Decision-Making and Access to Justice in Environmental Matters (Aarhus Convention, 1998). The Protocol on Water and Health was signed at London 1999. At the very beginning of new Millennium at 5th Ministerial Conference "Environment for Europe" held 21–23 May 2003 in Kiev". Several new treaties were signed, aimed at more detailed regulation of the issues comprised by the mentioned conventions. Those are the Protocol on Strategic Environmental Assessment (SEA Protocol), the Protocol on Civil Liability and Compensation for Damage Caused by Transboundary Effects of Industrial Accidents on Transboundary Waters and the Protocol on Pollutant Release and Transfer Registers (PRTR Protocol); UNDP (2007), pp. 41–42.

[22] Depending on the criteria/definition applied, today over 20 international (transboundary/shared) river basins (sub-basins) can be counted, including four lake basins. A number of them are sub-basins of the Sava River Basin, which itself is one of the sub-basins of the Danube Basin.

[23] For some more details incuding maps, see UNEP/GRID-Arendal (2007), pp. 46–48. An overall map of the SEE river basins is attached as the Appendix to this paper.

4 The EU Context

Taking into consideration that all Balkan countries declared their commitment to participate to the European integration process with the ultimate goal of becoming members of the European Union, the entire body of the *acquis communautaire* has to be transposed into their national legal systems and evidence of the implementation and enforcement of transposed legislation have to be provided to the EU. Compliance with the entire EU legal system is a condition *sine qua non* for the association of new countries to the EU. The obligation of transposition, implementation and enforcement of the EU requirements in a defined period of time is set in the Stabilization and Association Agreements (SAA) concluded between EU and the candidate countries. A large part of the Community *acquis* deals with environmental protection, which the waters of EU and neighbouring countries are part of. The new water management paradigm, based on the river basin management plan, has been designed and set in the Water Framework Directive – WFD (2000/60/EC) and in the subsequently adopted directives on management of floods (2007/60/EC) and groundwater (2006/118/EC).[24]

The Accession to the EU implies the candidate countries to accept all those international treaties to which the EU is a party. In case of the mentioned UNECE conventions, this means that SEE countries must comply with them

---

[24] There is a much larger number of the EU legal instruments, applicable to waters. Some of the relevant directives are: Directive 2003/35/EC of European Parliament and of the Council of 26 May 2003 providing for public participation in respect of drawing up certain plans and programmes relating to the environment and amending with regard to public participation and access to justice Council Directives 85/337/EEC and 96/61/EC; Directive 2001/42/EC of the European Parliament and of the Council of 27 June 2001 on the assessment of the effects of certain plans and programmes on the environment (SEA Directive); Council Directive 96/EC of 24 September 1997 concerning Integrated Pollution Prevention and Control (IPPC Directive); Council Directive 97/11/EC of 3 March 1997 amending Directive 85/3337/EEC on the assessment of the effects of certain public and private projects on the environment (EIA Directive); Council Directive 91/616/EEC of 12 December 1991 concerning the protection of waters against pollution caused by nitrates from agricultural sources; Council Directive 91/271/EEC of 21 May 1991 concerning urban wastewater treatment; Council Directive 80/778/EEC of 15 July 1980 relating to the quality of water intended for human consumption; Council Directive 86/280/EEC of 12 June 1986 on limit values and quality objectives for discharges of certain dangerous substances included in the List I of the Annex to the Directive 76/464/EEC; Council Directive 84/491/EEC of 9 October 1984 on limit values and quality objectives for discharges of hexaclorocyclohexane; Council Directive 84/156/EEC of 8 March 1984 on limit values and quality objectives for mercury discharges by sectors other than the chlor-alcaly electrolysis industry; Council Directive 83/513/EEC of 26 September 1983 on limit values and quality objectives for cadmium discharges; Council Directive 82/176/EEC of 22 March 1982 on limit values and quality objectives for mercury discharges by the chlor-alcali electrolysis industry; Council Directive 76/464/EEC of 4 May 1976 on pollution caused by certain dangerous substances into the aquatic environment of the Community. A number of those directives were and shall be gradually (e.g. 2007, 2013) replaced by the legal provisions adopted in the framework of a new (so called "shifted") paradigm, established by the Water Framework Directive.

even if they are not the parties, since they are part of EU law. For example, in 2004 and 2005, Serbia successfully transposed into its legal system the EU Directives on Strategic Environmental Assessment–SEA, Environmental Impact Assessment – EIA, and Integrated Pollution Prevention and Control – IPPC, (through the adoption of three respective laws and a great deal of secondary legislation in the form of governmental decrees and ministerial regulations). All the related requirements of the Espoo Convention and the SEA Protocol were implemented through this legislation. A recent (2006) detailed investigation on the implementation on the Aarhus Convention requirements for public participation in decision making and access to justice in environmental matters in the procedures set in the new legislation on SEA, EIA and IPPC has shown full compliance of Serbia with the requirements of the Aarhus Convention too, to which Serbia was not a party.[25]

This does not mean that SEE countries should not be parties to the international treaties to which the EU is a party. There are serious reasons for all those countries to be parties to the UNECE environmental conventions, which play an important role in the protection and management of the global environment, strengthen environmental policies and commitments at the national level and provide the frameworks for trans-boundary co-operation in dealing with environmental degradation. UNECE and EU policy requirements were concisely reaffirmed as the basis of co-operation by the member states of the OSCE Parliamentary Assembly, to which all the SEE countries are members, since 2005.[26]

The same is true for accession/ratification to the UN Convention on the Law of Non-Navigational Uses of International Watercourses. Co-operation on the principles and in the frameworks of those international environmental and water law instruments can only yield benefits, and contribute to the stability and security to the SEE sub-region.

The ability to comply with *acquis* requirements, in regard of water management has proven to depend on the ability of each state to successfully change its national water management system (inherited from socialist era, except in Greece and Turkey) and on the willingness to accept the new water management paradigm, inherent to which is an active co-operation with neighbouring/river basin countries. Most of the SEE countries are on their way to develop in a feasible time new and modern water management systems and to replace old and outdated water-economy management patterns. In searching for better system solutions, all of them, except Turkey, enacted new water laws, or are on their way to adopt new water legislation (for details, see the Table).

---

[25] UNEP/GRID-Arendal (2007).
[26] OSCE Parliamentary Assembly (2005).

Table Water Laws of the SEE countries

Country		The title	Enacted
Albania		Water Law	1996
		Law on Protection of International Lakes	2003
BiH	Federation BiH	Water Law	2006
	Republic of Srpska	Water Law	2006
Bulgaria		Water Act	1999
Croatia		Water Law	2009
Greece		Water Law	2003
Macedonia		Water Law	1998
Montenegro		Water Law	2007
Romania		Water Act	1996
Serbia		Water Law	2010
Kosovo		Water Law	2004
Turkey		Water Law	1926
		Law on establishment of Gen. Dir. of State Hydraulic Works	1953
		Groundwater Law	1960

5 The Evolution of the New Legal Regimes

When developing new (bilateral) legal regimes for shared water resources, new states must rely on all those (mentioned and other not mentioned) international policy and legal requirements, applicable to the SEE sub-region. It is not likely that any contemporary project on international waters prepared unilaterally, or neglecting certain vital principles (like public participation in decision making) could be successful.[27]

The new SEE states have acceded to the Danube River Protection Convention[28] and concluded several bilateral agreements concerning cooperation with regard to their

---

[27] Quite recent experience from Montenegro can serve as an example when NGOs campaign was successful in protection of the Tara River from the construction of a large hydroelectric power plant (*Buk Bijela*) in the Sava River Basin. See Natasa Djereg (2010); Lazarela Kalezic (2010). Actually, this case indicates something more. Namely, besides of the Montenegrin and the Republic of Srpska (one of two BiH Entities) authorities, such project requires involvement of other basin authorities, i.e. of BiH and Serbia, as well as the Sava Commission. An equally important stakeholder, because of protection of the Tara River, is UNESCO. It seems that only a (sub-)regional, river basin approach (defined in a water treaty), based on the well know and clear principles of international water and environmental law, can provide sound fundaments for undertaking of new hydro-electric power developments in this case. Quite recently, attempts for implementation of certain projects for hydrolectric structures in BiH and Croatia faced also strong oposition from the general public, and especially environmental NGOs. For more information see e.g. OSCE (2007).

[28] See http://www.icpdr.org/icpdr-pages/contracting_parties.htm. The web page last visited 2 February 2010.

shared waters.[29] Macedonia and Albania have good cooperation on shared waters.[30] Collaboration is visible between Montenegro and Albania,[31] as well as between Albania, Greece and Macedonia,[32] and Bulgaria, Moldova, Romania and Ukraine.[33]

Nevertheless, the most remarkable regional achievement is, without any doubt, the ratification of the Framework Agreement of the Sava River Basin – FASRB (2002), between the Sava River Basin countries, i.e. BiH, Croatia, Federal Republic of Yugoslavia[34] and Slovenia). This agreement, the first one at

---

[29] The Agreement on Water Economy Relations between the Government of the Republic of Croatia and the Government of the Republic of Hungary was signed in 1994 ("Official Gazette of the Republic of Croatia (International Treaties)", No. 10/94); The Contract between the Goverment of the Republic of Croatia and the Government of Bosnia and Herzegovina on Regulation of Water Economy Relations was signed 1996 ("Official Gazette of the Republic of Croatia (International Treaties)", No. 10/96); The Contract between the Goverment of the Republic of Croatia and the Government of the Republic of Slovenia on Regulation of Water Economy Relations was signed 1997 ("Official Gazette of the Republic of Croatia (International Treaties)", No. 10/97); The Contract between the Goverment of the Republic of Croatia and the Government of the Republic of Montenegro on Mutual Relations in the Field of Water Management was signed 1998 ("Official Gazette of the Republic of Croatia (International Treaties)", No. 1/98).

[30] For example, the Agreement between the Council of Ministers of the Republic of Albania and the Government of the Republic of Macedonia for the Protection and Sustainable Development of the Lake Ohrid and its Watershed, was signed on 17 June 2004 at Skopje.

[31] Concerning e.g. the Skadar/Shkoder Lake in terms of The Memorandum of Understanding for Cooperation in the Field of Environment Protection and Sustainable Development Principle Implementation on the Skadar/Shkoder Lake between the Ministry of Environment of the Republic of Albania and the Ministry of Environment and Physical Planning of the Republic of Montenegro, signed in Podgorica, Montenegro, on 9 May 2003. However, the activities of Albania on the development of the *Ashta* hydropower production plant on the Drim/Drini River, without previously having completed the procedure for environmental impact assessment of the project on the environment of neighbouring Montenegro, contrary to the Espoo Convention which the both States are the Parties of Lazarela Kalezic (2010), might become one of the issues for concern from the security standpoint.

[32] In the case of the Prespa Lakes Basin, a new multilateral agreement concerning the protection and sustainable development of the Prespa Park, was signed recently. This move of the Prespa Lakes litoral countries towards signing an agreement was based on the remarkable results of transboundary cooperation in the last decade, developed in the frameworks of the Declaration on the Creation of the Prespa Park and the Environmental Protection and Sustainable Development of the Prespa Lakes and their Surroundings, signed on 2 February 2000 at Agios Germanos, Greece. For more details see *infra*, Note 46.

[33] Mentioned should be e.g.: the Agreement between the Government of the Republic of Moldova and the Government of Ukraine on Joint Use and Protection of Transboundary Waters, signed on 23 November 1994 at Chisinau; the similar agreement between Romania and Ukraine concerning relations between two governments in regards of shared waters signed on 30 October 1997; the Agreement between the Ministry of Environment and Territorial Planning of the Republic of Moldova, the Ministry of Waters, Forests and Environmental Protection of Romania and the Ministry of the Environment and Natural Resources of Ukraine on the Cooperation in the Zone of the Danube Delta and Lower Prut Nature Protected Areas, signed in Bucharest, on 5 June 2000. Bogdanovic (2005), p. 82.

[34] I.e. Serbia as a successor of the Community of States Serbia and Montenegro. Montenegro, as the fifth and most upstream country in the Sava River Basin has been still awaited to accede to the FASRB.

the sub-regional level after the conclusion of the Dayton-Paris Peace Agreement (1995) is based on the equal desire of its Parties "to co-operate in a constructive and mutually beneficial manner". At the same time, a protocol was signed on the regulation of the navigation regime on the Sava River and its tributaries, and the Sava Commission, as the FASRB implementing body with its international Secretariat in Zagreb, was established. New protocols, that should "fill" the FASRB frameworks and enable its implementation (with regard to e.g. transboundary impacts, protection against floods, groundwater, erosion, ice hazards, draughts and water shortages, water use, protection of aquatic eco-systems, prevention of water pollution caused by navigation and so on) are still in the drafting phase.[35]

6 Open Fields for Co-operation

Apart from the need to revise old (not yet replaced) water treaties discussed above, existing in the SEE, there is a number of issues, mostly related to bilateral relations on shared waters, that should be regulated between BiH and Montenegro,[36] Albania and Montenegro,[37] BiH and Serbia,[38] BiH and Croatia,[39] Croatia and Serbia,[40] Macedonia and Serbia,[41] Greece and Macedonia,[42] Bulgaria and Greece,[43] Bulgaria, Greece and Turkey.[44]

A positive development of co-operation based on an authoritative soft law instrument – the Prespa Lakes Declaration[45] may be expected to lead to significantly improved and enhanced cooperation based on the new agreement signed between Albania, Greece, Macedonia and the Commission of the European

---

[35] For more details see Slavko Bogdanovic (2004); International Sava Basin Commission (2010).

[36] Use of the Bilecko Lake waters or certain developments in the upper parts of the Drina River Basin.

[37] An agreement applicable on the Skadar/Shkoder Lake, binding for the Lake littoral states, would be an expected outcome of on-going cooperation based on existing Memorandum of Understanding of 2003 (see *supra*, Note 31).

[38] Inter-state border on the Drina River, and particularly development of exploitation of the Drina River hydropotential.

[39] The Neretva and Trebisnjica River Basin. For more details see:http://www.evd.nl/zoeken/showbouwsteen.asp?bstnum=131781&location; Web site last visited 23 November 2009.

[40] Inter-state border on the Danube River and colaboration of water management authorities of the two countries.

[41] The Pcinja River Basin.

[42] The Vardar/Axios River Basin; the Dojran Lake.

[43] The Struma/Strimonium, the Maritsa and the Mesta/Nestos River Basins.

[44] The Tundza, Arda and Ergene/Evros River Basins.

[45] See *supra*, Note 32.

Communities at Pyli, Greece, on 2 February 2010.[46] Similarly, new river/river basin treaties might be necessary for a better management of the Drim/Drini River Basin[47] as well as for the waters of the Tisa River Basin.[48]

Relations between Serbia and Kosovo regarding the waters crossing their borders should be regulated. The population living in Serbia downstream the rivers flowing from and through Kosovo, have a clear interest for that, at least in terms of up-stream control of pollution of those waters.

Necessity and potential for new developments of hydropower production in the SEE region in the near future are huge. The European Bank for Reconstruction and Development (EBRD) data of 2006 show estimations that (approximately) 30% of hydro potential in Romania, Bosnia and Herzegovina, and Serbia and Montenegro,[49] 60% in Bulgaria, 50% in Albania and 25% in Macedonia has not been used yet.[50] Most of these reserves could be used through development of trans-boundary watercourses. Such development activities imply cooperation between interested states and regulation of their relations by international treaties (bilateral and in certain cases trilateral or even multilateral).

7 A Glance at Traditional Water Management Patterns

When looking from a contemporary perspective to the national water management systems in the SEE (managing national waters and complying with old water treaties, where they exist, and trying to comply with the new EU water management paradigm) and having in mind the modern dynamic development of the concepts of

---

[46] The agreement contains specific commitments but also opportunities for the natural environment and human activities in the Prespes region, and will introduce standing cooperation structures that are deemed necessary to confront the wetlands' problems and also for the region's sustainable development.

The agreement for the Park that straddles the borders of all three countries was signed by Greek Environment Minister Tina Birbili and her Albanian and FYROM counterparts, Fatmir Mediu and Nexhati Jakupi, as well as European Commissioner for the environment Stavros Dimas. It dates back to the pledge made in a Joint Communique on 2 February, 2000 by the prime ministers of the three countries at that time to cooperate in protecting the Prespes ecosystem. See: http://www.ana.gr/anaweb/user/showplain?maindoc = 8374584&maindocimg = 8374063&service = 144. Web page last visited 2 February 2010.

[47] Shared by Macedonia, Albania, Montenegro and Kosovo.

[48] In accordance with commitment of the signatories, expressed in the Paragraph 9 of the Towards a River Basin Management Plan for the Tisza River Supporting Sustainable Development of the Region Memorandum of Understanding, signed by Tisa riparian states (Serbia and other upstream countries, three of which are EU member countries, i.e. Hungary, Romania and Slovakia, and Ukraine) at Vienna on 13 December 2004.

[49] The joint data for two countries, before 2006 splitting in two independent states.

[50] See UNEP/GRID-Arendal (2007), p. 55.

good water governance[51] (and subsequently established new managerial policy requirements and new legal frameworks), it seems easy to find out that inherited water economy systems in the Balkans:

- Are based on (in many cases strongly) centralized ("top down") approach
- Are not allowing public involvement in decision making
- Are rarely and rather declaratively dealing with environmental protection issues
- Do not foresee proper instruments for implementation of declared commitments
- Are taking the water resources in a piece-meal manner in many senses, without integrated, river basin and eco-system approach, in spite of often repeated declarations on the contrary

Besides, in certain cases the old institutional arrangements established for the implementation of international water treaties in the region, as well as their work and achievements/lack of results, were/are safely situated far from the public insights.

Intensive (and increasing) inter-action between legal requirements coming from the international level and national legislative level, involve today as never before different national authorities competent for different aspects regarding the management of the same (trans-boundary) water resource. National authorities competent for agriculture, forestry and water economy, traditionally the sectoral hub for all aspects of water economy, seems to have no more ability and capacity for coping with concurrent aspects of the management of national or international waters.

In others words, other governmental sectors, competent for e.g. protection of water quality, preservation and protection of aquatic ecosystems, providing water for human consumption, industrial and municipal use, navigation, hydropower production, protection against various detrimental effects from waters/lack of water (originated from climate changes or from anthropogenic factors) have segmented competence. Substantive changes of national water management systems should be aimed at achieving a high degree of integration of the water governance structure at all levels (international, national, regional and municipal). This may encompass the replacement of outdated management techniques and the development of suitable and supportive ways of co-operation among all the competent public authorities, the business sector and the civil society.[52]

---

[51] For more details on legal aspects of good water governance see Allan and Wouters (2004).
[52] UNEP/GRID-Arendal (2007), p. 53.

8 Remarks on Certain Security Aspects

There are examples confirming that in the SEE region water resources or hydraulic structures themselves were used during WWII and also in recent times as military tools or targets, as political tools or terrorist targets.[53] Obviously, in times of armed conflicts, the SEE region has been prone to such use (i.e. destruction/spoiling) of waters and hydraulic structures.

Today, in times of peace, certain open issues could serve as the basis for new disputes and political tensions. Such could be the case of those river basins in the region not covered yet by any water treaty. Furthermore, water resources could be used in some cases as a political tool (in case of non-settled inter-state borders on the Danube River, between Serbia and Croatia and on the Drina River, between Serbia and Bosnia and Herzegovina). Development projects (e.g. hydraulic plants construction, like in case of *Buk Bijela* and *Ashta*), unilateral implementation of certain projects that could have undesirable trans-boundary impacts, breaching of existing legal regimes (e.g. disrespect of Espoo Convention, that might lead in a dispute similar to the *Kanal Bistroe* dispute), could also be source of such disputes.

The most serious situations in SEE could arise in case of the suspension of existing (and highly coherent and consistent) legal regimes applicable to the projects with a trans-boundary impact (national, regional UNECE, EU) established for controlling trans-boundary impacts. A multilateral treaty signed recently – Multilateral Agreement among the Countries of South-Eastern Europe for the Implementation of the Convention on Environmental Impact Assessment in a Trans-boundary Context seems to contain such a potential.[54] According to Article 6 Paragraph 2 of this Agreement, if the Parties agree, for joint (trans-boundary) projects/activities (neither criteria nor specification of these project/activities were set in the Agreement) EIA procedures (as requested by national legislation or Espoo Convention) shall not apply.[55]

Certainly, it does not seem realistic at all that any kind of armed conflict could evolve in the SEE region as a result of non settled water issues.[56] Changed

---

[53] For details see Peter Gleik (2009), pp. 11, 12 and 14. To the Gleick's List of Water Conflict Chronology (containing various acts done in Yugoslavia 1993, 1999, Kosovo 1998, 1999 and Macedonia 2001) should be added the destruction by German XIII SS *Handzar* Division, in the beginning of 1944, of the pump station on the mouth of the River Bosut (close to its confluence with the River Sava) in Yugoslavia (Srem, Vojvodina) aimed at flooding a huge flat forested area (*Bosutske sume*) where the civilian shelters and camps of partisan military forces were established. Under the Gleick's typology this act could be put into the categories of "Military Tool" and "Military Target".

[54] This treaty was signed at the Fourth Meeting of the Parties to the Espoo Convention, at Bucharest, held 19–21 May 2008.

[55] For details see Natasa Djereg (2010).

[56] For considerations concerning "wars on water" particularly in the context of the Tigris and Eupohrates River Basin countries relations and acceptance of the UN Convention on the Law of Non-Navigational Uses of International watercources (New York, 1997) see e.g. Scherk et al. (1998).

perception of state sovereignty (over natural resources) in terms of acceptance of broad global and European legal frameworks (established by UN, CSCE/OSCE, UNECE and EU) based on reasonable and equitable use of waters (and share of benefits), on avoiding to cause significant harm to neighbouring states, and on respecting growing requirements for recognition and protection of human rights, could lead to a favourable environment for the establishment of the still missing adequate legal regimes through the conclusion of international treaties. Consequently, the focus would be on states' compliance with accepted binding duties. With the signature of international water treaties, the development based on the protection and the use of shared water resources would seem more sustainable.

9 Conclusion

In the contemporary globalizing and fast changing world, facing often unexpected occurrences caused by climate change and searching to achieve sustainable development and fulfil Millennium Development Goals, the need for a profound change of the traditional national (rather crumbled than integrated, but still centralized) water management systems is evident. Such an unavoidable change must lead towards the acceptance and implementation of new conceptual global, UNECE and EU policy requirements, which are reaffirmed as the basis of co-operation by the member states of the OSCE Parliamentary Assembly.

The action should be directed towards the full embrace of the shifted water management paradigm, i.e. through joining indicated multilateral international treaties, and changing of existing international and national legal instruments or through the development of new ones that would better express the current European integrated/sustainable water resources management trends.

Finally, having in mind a considerably composite network of (global, regional, EU, sub-regional, bilateral, river/river basin) legal rules, and considering in particular the fast growing policy and soft-law requirements, applicable on SEE water resources, being they trans-boundary or national, fresh, surface or underground, it is difficult to imagine successful operation of any national water management system if not based on the applicable international treaties. The acceptance of the UN/UNECE relevant water and environmental treaties and the full compliance with them as well as with the *acquis communautaire* and the regulation of neighbouring relations with regard to shared waters shall provide conditions for avoiding water disputes, leaving outdated management practices behind and introducing adequate water management systems capable to face contemporary challenges.

Acknowledgment I wish to express my gratitude to UNEP/GRID-Arendal for permission to use the SEE River Basins map attached in the Appendix to this paper.

Appendix

The Map of the SEE River Basins

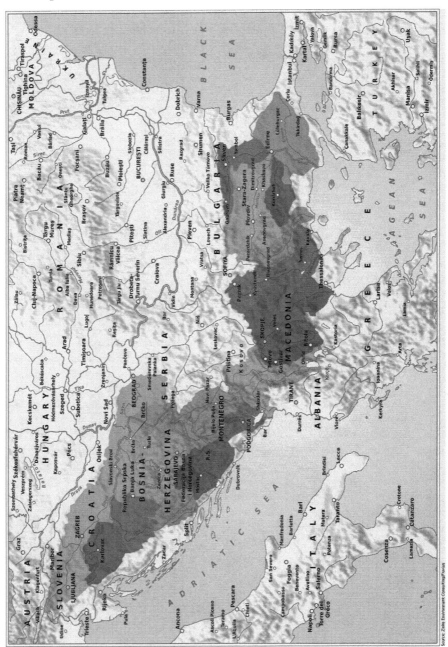

References

Allan A, Wouters P (2004). What role for water law in the emerging "good governance" debate? 15 Water Law, pp. 85–89

Birnie P, Boyle A (2002) International law & the environment. Oxford University Press, New York

Bogdanovic S (2004) Good governance of the Sava Basin – legal realities and practical prospects; Lawtext Publishing Ltd. J Water Law 15(WL3/04):93–99

Bogdanovic S (2005) Legal aspects of trans-boundary water management in the Danube Basin. Arch Hydrobiol (Suppl. 158):1–2; Large Rivers Vol. 16, No. 1–1; E. Schweizerbart'sche Verlagsbuchhandlung; Stutgart; pp. 59–93

Bourne C (2001) The international law association's contribution to international water resources law. In: Bogdanovic S (ed) International law on water resources – contribution of the international law association (1954–2000). Kluwer Law International, London, The Hague, Boston

Council of Europe (2006) Action plan for conservation of sturgeons (*Acipenseridae*) in the Danube River Basin; Strasbourg Cedex

Djereg N. EIA in the trans-boundary context of SEE; in this publication

Gleich P (2009) Water Conflict Chronology. http://www.worldwater.org.conflict.html. The web page last visited 22 Nov 2009

International Sava Basin Commission (2010) Basic docs. http://www.savacommission.org/basic_docs. The web page last visited 02 Feb 2010

Kalezic L EIA in a trans-boundary context in Montenegro; in this publication

Lammers GJ (2008) Potential effects from the non-entry into force of the UN watercourses convention; Seminar: the UN watercourse convention: legacy, prospects and value for the realization of international policy goals; 2008 World Water Week, Stockholm, Sweden, 17–23 August 2008. http://www.worldwaterweek.org/downloads/UNWC_Seminar_Lammers%5B1%5D.pdf. The web site last visited 11 Jan 2009

OSCE (2005) OSCE Parliamentary assembly: Washington declaration: 30 years since Helsinki: challenges ahead. Washington, DC 01-05 July 2005

OSCE (2007) Briefing paper: a look at water management in Bosnia and Herzegovina; Second preparatory conference to the 15th OSCE economic and environmental forum, Zaragoza, Spain, 12–13 March 2007. http://www.OSCE.org/documents/eea/2007/03/23783_en.pdf. The web site last visited 05 Feb 2010

Sherk GW, Wouters P, Rochford S Water wars in the near future? Reconciling competing claims for the world's diminishing freshwater resources – the challenge of the next millennium; The CEPMLP on-line Journal, Vol. 3, No. 2. 1998 http://webworld.unesco.org/water/wwap/pccp/cd/pdf/educational_tools/course_modules/reference_documents/issues/waterwarsinthenearfuture.pdf

UNDP (2007) Environmental policy in South Eastern Europe; conference "Environment for Europe". UNDP Podgorica, Belgrade

UNEP (2007) UNEP/GRID-Arendal. Balkan Vital Graphics, Arendal

WWF calls on the world's governments to join and implement the 1997 UN convention on the law of the non-navigational uses of international watercourses (UN watercourses convention). http://www.panda.org/what_we_do/how_we_work/policy/conventions/water_conventions/un_watercourses_convention/. The web page last visited 02 Feb 2010

Watts A (1999) The international law commission 1949–1998. The Treaties, Part II, Vol. Two. Oxford University Press, Oxford

Towards Environmental Security by Adapting the Energy Sector: Summary of Strategies and Opportunities for Technology Transfer and Cooperation in the SEE Region*

Vanessa Peña and Brandon Petelin

Abstract Within the SEE region, there are several opportunities for energy sector adaptation in response to environmental, energy, and climate security concerns. These latter inter-dependent security issues have similar over-arching aims that could contribute to an effective environmental regime. This paper surveys five fundamental criteria for an effective environmental regime within the context of renewable energy and energy efficiency policy in the SEE region. These five criteria are: (1) effective policies, laws, and regulations based on desired objectives (the legal base-line aspect); (2) an embedded desire to pursue environmental objectives, which are most likely based on incentives (the environmental culture aspect); (3) strong national institutions that can pursue objectives and enforce measures when necessary (the institutional aspect); (4) national capacity, which to a large extent depends on effective technology transfer (the capacity building aspect); and (5) given the international impact of environmental issues, regional cooperation, as well as, support from international institutions (the regional and international cooperation aspect). The countries in the SEE region have made significant progress in developing renewable energy and energy efficiency policies, programmes, and projects, but there remain regulatory, economic, and institutional barriers to effective energy adaptation. As a result, important opportunities exist in the areas of technology transfer and capacity building as well as regional collaboration.

---

*Please note the viewpoints expressed are solely those of the authors and do not represent those of the Institute for Defense Analyses or the US Department of Commerce.

V. Peña (✉)
Science and Technology Policy Institute, Institute for Defense Analyses,
1899 Pennsylvania Avenue NW, Washington, DC 20006, USA
e-mail: vpena@ida.org

B. Petelin
U.S. Department of Commerce, International Trade Administration,
1401 Constitution Avenue NW, Washington, DC 20230, USA
e-mail: brandon.petelin@trade.gov

Keywords Environmental security • Energy security • Climate security • Renewable energy • Energy efficiency • Kyoto Protocol • Energy Community • *Acquis communautaire* • Technology transfer • Clean Development Mechanism (CDM) • Regional cooperation

1 Introduction

Over the years, conflict in the South Eastern European (SEE) region has resulted in slow growth, low living standards, and increased environmental stress. Solutions to the region's environmental problems and the achievement of greater environmental security have, thus, become central components of sub-regional and trans-border cooperation. In particular, effective environmental protection management and understanding the role of resource exploitation across national borders can contribute to conflict prevention and peace building efforts (UNEP 2009). Environmental protection and security goals are inherently recognized as being inter-dependent on national security, energy and natural resources, climate change, social and economic stability, and conflict prevention. For example, energy security has became a principle concern in early 2009 as Russia disputed gas prices and cut off supply to Ukraine, a major transit country for much of the SEE region's gas imports, with the worst affected SEE countries being Bulgaria, Croatia, and Serbia (Crooks 2009). Moreover, there is pressure building in the SEE region, as well as in the global community, to respond to calls for climate security.

Thus, within the SEE region there are several opportunities for energy sector adaptation in response to environmental, energy and climate security concerns. Despite the differences among SEE countries, which stem from varying stages of development and differing ideas about effective economic and development strategies, certain opportunities within the energy sector, in renewable energy and energy efficiency, could generate meaningful results. While these opportunities are SEE specific, many of them require cooperation among a range of both internal and external stakeholders if the SEE region is going to effectively adapt in light of future energy challenges. While the supportive framework and policies to adapt in response to these concerns and the necessary motivations (e.g. increased global competitiveness within the region, less external energy dependency) to work collectively are in place, the lack of effective political leadership, institutional capacity, and a comprehensive environmental regime hinder efforts for adapting the energy sector (Crooks 2009).

Generally, there are several requirements for a workable and effective environmental regime (OECD 2008). Most notably, such a regime requires at least five elements: (1) effective policies, laws and regulations based on desired objectives (the legal base-line aspect); (2) an embedded desire to pursue environmental objectives, which are most likely based on incentives (the environmental culture aspect); (3) strong national institutions that can pursue objectives and enforce measures when necessary (the institutional aspect); (4) national capacity, which to a large extent depends on effective technology transfer (the capacity building aspect); and

(5) given the international impact of environmental issues, regional cooperation, as well as support from international institutions (the regional and international cooperation aspect) (OECD 2008). Therefore, in this paper, we will survey each of these elements with respect to the SEE region. First, we will briefly review the relevant energy policies (Sect. 2) and economic incentives (Sect. 3) in place to achieve renewable energy and energy efficiency within the SEE region. We will then specifically consider opportunities for technology transfer and capacity building (Sect. 4). Finally, we will discuss the opportunities for institutional development regional cooperation, and international support (Sect. 5.2).

2 Regulatory Framework: The Legal Baseline

There are a variety of important international and national policies that provide an enabling environment and supportive framework for the implementation of renewable energy and energy efficiency policies, programmes, and projects within the SEE region. For reference, the SEE countries of interest throughout this discussion include: Albania, Austria, Bosnia and Herzegovina, Bulgaria, Croatia, Hungary, Montenegro, Romania, Serbia, Slovakia, Slovenia, The Former Yugoslav Republic of Macedonia, Turkey, and the United Nations Interim Administration Mission in Kosovo (UNMIK).

At the international level, the Kyoto Protocol under the United Nations Framework Convention on Climate Change (UNFCCC) and the Treaty Establishing the Energy Community (2005) are the two major agreements that have driven national energy-related policies and actions. Most SEE countries have acceded to or ratified the Kyoto Protocol and several countries are considered as an Annex I Party to the Protocol, which indicates countries committed to national or joint emission reduction targets. Accession, ratification, and entry into force dates for specific SEE countries are shown in Table 1 (Annex 1 countries are indicated by an asterisk). The Kyoto Protocol provides three major mechanisms – Emissions Trading, the Clean Development Mechanism (CDM), and Joint Implementation (JI) – that provide significant opportunities and incentives to drive investments in renewable energy and energy efficiency technologies from Annex I countries to countries designated as developing countries or economies in transition. Specifically, opportunities for technology transfer and capacity building through CDM activities are discussed in Sect. 4.

Another significant international treaty specific to the European region is the Treaty Establishing the Energy Community, which was signed in October 2005 (entered into force in July 2006) by the European Community and Albania, Bosnia and Herzegovina, Bulgaria, Croatia, Montenegro, the Former Yugoslav Republic of Macedonia, Romania, Serbia and UNMIK on behalf of Kosovo ('Contracting Parties') (Energy Community Treaty 2005). At present the Treaty has 7 Parties and 14 Members States of the EU have acquired the status of 'Participants' to the Treaty. The Treaty is aimed at ensuring appropriate integration of the SEE's regional energy market into the EU's internal energy market and enhancing the security of energy supply (Energy Community 2009c). Contracting Parties to the treaty are required to

Table 1 Kyoto Protocol accession and ratification in SEE countries

Country	Accession	Ratification	Entry into force (% of emissions for Annex I Parties)
Albania	April 2005	2007	June 2005
Austria*	–	May 2002	February 2005 (0.4%)
Bosnia and Herzegovina	April 2007	–	July 2007
Bulgaria*	–	August 2002	February 2005 (0.6%)
Croatia*	–	May 2007	August 2007
Hungary*	August 2002	–	February 2005 (0.5%)
Montenegro	June 2007	–	September 2007
Romania*	–	March 2001	February 2005 (1.2%)
Serbia	October 2007	–	January 2008
Slovakia*	–	May 2002	February 2005 (0.4%)
Slovenia*	–	August 2002	February 2005
The Former Yugoslav Republic of Macedonia	November 2004	–	February 2005
Turkey*	May 2009	–	August 2009
United Nations Interim Administration Mission in Kosovo (UNMIK)	–	–	–

Data from UNFCCC 2009
*Indicates an Annex I Party to the UNFCCC (UNFCCC 2009b)

implement the *acquis communautaire* and develop national implementation plans within a fixed time frame in preparation for integration into the EU. This includes the transposition of two EU Directives focused on renewables – Directive 2001/77/CE promoting electricity produced from renewable energy sources and Directive 2003/30/EC promoting the use of bio-fuels and other renewable fuels for transport – in addition to the *acquis* on electricity, gas, environment, and competition. The requirements in relation to these Directives are addressed by the Contracting Parties within their Implementation Plans. However, the recent passage of Directive 2009/28/EC on the promotion of the use of energy from renewable sources amends and subsequently repeals the Directives 2001/77/EC and 2003/30/EC effective in 2012 (Energy Community 2009a). Thus, Energy Community members are currently studying the prospects for adopting the new EU Renewable Energy Directive through the Renewable Energy Task Force, which was recently established after the 5th Ministerial Council meeting in 2008 (Energy Community 2009f).

Furthermore, the Energy Community's Energy Efficiency Task Force (EETF) was created in December 2007, and its mandate was recently extended by the Ministerial Council until the end of 2010 (Energy Community 2009h). The activities of the EETF are to help finalize the National Energy Efficiency Action Plans (NEEAPs), start the implementation and monitoring of the NEEAPs, and implement the Awareness Raising Campaign. Previous work of the EETF included the identification and preparation towards the implementation of three EU Directives on energy efficiency – Directive 2006/32/EC on energy end-use efficiency and energy services, Directive 2002/91/EC on the energy performance of buildings, and Directive 92/75/EEC on the indication by labelling and standard product information of the consumption of energy and other resources by household appliances. Most recently, in December 2009, the Ministerial Council of the Energy Community adopted the decision No. 2009/02/MC-EnC on the implementation of Directive 2006/32/EC on energy end-use efficiency and energy services by the Contracting Parties to the Energy Community (Energy Community 2009b).

On the other hand, at the national level, policies or programmes related to renewable energy and energy efficiency have been implemented by most SEE countries. Table 2 provides a summary of energy laws, energy strategy documents, renewable energy legal frameworks, and energy efficiency legislation or programmes for chosen SEE countries. The majority of the SEE countries shown in Table 1 have an Energy Law and an Energy Strategy document describing the basic principles for the energy sector. However, some of these laws or documents sometimes address renewable energy and energy efficiency rather vaguely. For example, some of these documents generally encourage the 'promotion of the use of renewable energy sources' and a 'more efficient use of energy' rather than laying out specific supportive frameworks or programmes that are needed to achieve these goals. To make this distinction, SEE countries that have adopted a specific law or programme for renewable energy or energy efficiency are also noted in Table 2.

In many cases, the formulation of national renewable energy policies and energy efficiency programmes is primarily driven by the obligations of the Contracting Parties to the Energy Community Treaty. For example, many policies aimed at promoting electricity production and access to the grid system adopt the renewable energy and energy efficiency legal frameworks established in the Treaty, such as Austria's Green Electricity Act, the drafting of Chapter VII Generation from Renewable Energy Sources within Montenegro's Energy Law, and the transposition of Directive 2001/77/EC within the Former Yugoslav Republic of Macedonia's various rulebooks. Further, Albania, as of October 2009, is drafting a new law for the promotion of renewable energy sources (Tugu 2009). The UNMIK has not yet addressed a legal framework for renewable energy sources.

In terms of energy efficiency, Albania has adopted a draft energy efficiency law and objectives for energy efficiency are described in the National Strategy of Energy Action Plan (NAE 2005). Hungary and Bulgaria, through its Energy and Energy Efficiency Law of 1999, have implemented energy efficiency programmes included in the 2001 Energy Charter Protocol on Energy Efficiency and Related Environmental Aspects (PEEREA) (*see* SEEnergy 2009). Moreover, Croatia, Montenegro, Serbia, and the Former Yugoslav Republic of Macedonia have begun energy savings projects in public buildings (Energy Community 2009g). Bosnia and Herzegovina and the UNMIK have yet to sufficiently address energy efficiency through legislation or conservation programmes. For example, previous energy efficiency activities within the UNMIK have focused on public awareness and information campaigns. However, the Implementation Plan for the Energy Strategy of Kosovo for 2009–2011 suggests plans to develop a building energy efficiency project for the entire public sector by 2013 and implement more rigorous regulations for new buildings by 2011 (SEEnergy 2009; ACCK 2008).

Table 2 Summary of energy related legal frameworks and strategies

Country	Energy law	Energy strategy document	Renewable energy legal framework	Energy efficiency law or conservation program
Albania	No	Yes	Draft (as of Oct 2009)[b]	Yes
Austria[*]	No	Yes	Yes (Electricity Law)[c]	Yes
Bosnia and Herzegovina	No	Yes	Yes[d]	No
Bulgaria[*]	Yes	Yes	Yes[e]	Yes
Croatia	Yes	Yes	Yes[f]	Yes[l]
Hungary[*]	Yes	Yes	Yes[g]	Yes
Montenegro	Yes	Yes	Yes (Energy Law)[h]	Yes[l]
Romania	Yes[a]	Yes	Yes[i]	Yes
Serbia	Yes	Yes	Yes[j]	Yes[l]
The Former Yugoslav Republic of Macedonia	Yes	Yes	Yes (Energy Law, Rulebooks, and targets planned towards end 2009)[k]	Yes[l]
United Nations Interim Administration Mission in Kosovo (UNMIK)	Yes	Yes	No	No

Adapted from SEEnergy 2009
[*]Indicates an Annex I Party to the UNFCCC (UNFCCC 2009b)
[a] Law No. 13/2007 on Electricity (Bundesministerium für Umwelt, Naturschutz und Reaktorsicherheit 2009c)
[b] Energy Community, Recent Development on the Albanian RES Policies, Presentation to the 1st Meeting of the RETF, October 2009
[c] Green Electricity Act, Green Electricity Regulation 2006 (Bundesministerium für Umwelt, Naturschutz und Reaktorsicherheit 2007)
[d] Energy Community, Bosnia and Herzegovina - Renewable Energy Policy, Presentation to the 1st Meeting of the RETF 2009e
[e] Renewable and Alternative Energy Sources and Biofuels Act (Bundesministerium für Umwelt, Naturschutz und Reaktorsicherheit 2009a)
[f] Energy Community, Overview of Legislative and Institutional Framework in the Republic of Croatia Concerning Renewable Energy Sources, Presentation to the 1st Meeting of the RETF, October 2009
[g] Act No. LXXXVI of 2007 (Bundesministerium für Umwelt, Naturschutz und Reaktorsicherheit 2009b)
[h] Kovacevic I, Renewable Energy Policy in Montenegro, Presentation to the 1st Meeting of the RETF, October 2009
[i] Governmental Decision 1535/2003 approving the national strategy for the revaluation of renewable energy sources, Governmental Decision 958/2005 regarding the promotion of electricity production from renewable energy sources
[j] Stojadinovic D, Implementation of EU Acqui on Renewables, Presentation to the 1st Meeting of the RETF, October 2009
[k] Energy Community, Renewable Energy Task Force 1st Meeting, Presentation to the 1st Meeting of the RETF, October 2009
[l] Energy Community, Report to the 6th Ministerial Council on the Work Performed by the Energy Efficiency Task Force, May 2009

The progress made through transposition of the various EU Directives mentioned throughout the SEE countries is notable, particularly in Albania, Croatia, the Former Yugoslav Republic of Macedonia, and Serbia. But there are still regional and country-specific challenges in the implementation of goals aimed at energy and environmental security vis-à-vis clean energy production and energy conservation. The varied state of development for relevant energy policies may be due to differences in, or a lack of, commitment and necessary secondary legislation provided through a comprehensive regulatory framework (Energy Community 2009e). For example, there are a few countries that have implemented a national energy strategy that includes renewable energy electricity production targets, exceptions being Croatia and Romania, with plans for targets currently being assessed by the Former Yugoslav Republic of Macedonia and the UNMIK (Uzanov 2008; Karafilovska 2009; Raguzin 2009). Having clear targets for renewable energy production are important in order to give clarity and certainty to industry, to establish confidence among equipment suppliers and investors, and to provide clear benchmarks to measure progress. The Energy Community RETF is providing assistance to evaluate setting renewable energy targets for 2020 for the Contracting Parties. However, the analysis of associated costs for renewable energy projects may be hindered by the gaps and inconsistency in country data, the lack of reliable energy statistics, and the minimal data from examples of developed projects thus far (Energy Community 2009g).

Another impediment to the effective implementation of relevant energy policies is the availability of local resources, institutional capacity, and data to develop measures and assessments of renewable energy and energy efficiency potential as well as economic and social implications. Currently, many of the Contracting Parties to the Energy Community Treaty have not yet performed this type of assessment (Jelavic and Segon 2008). There is also a lack of appropriate technological development to ensure proper grid access to renewable electricity generation. Moreover, there are major obstacles in terms of modernization and rebuilding of the electricity grid in several SEE countries. Specifically, the damage to Bosnia and Herzegovina's energy infrastructure during the conflicts in the mid-1990s remains a major barrier to renewable energy development since governmental resources are focused on the repair of existing infrastructure rather than on new building projects or grid modernization (EBRD 2009a). Furthermore, the energy infrastructure in Serbia and Montenegro remains damaged due to conflicts with Kosovo in 1999. In this case, renewable energy development, other than the reconstruction of damaged hydropower facilities, is a low priority (EBRD 2009c). Finally, technological barriers may limit the access of relevant energy projects to the grid, such as in Bulgaria where the addition of new installations above 1,800 MW may put the reliability of the grid at risk (EBRD 2009b).

3 Economic Incentives: Creating an Environmental Culture

Several countries within the SEE region have been using renewable energy as a major source of energy consumption through hydropower electricity generation. The situations in Albania, which consumes close to all of its electricity from hydropower, and Bosnia and Herzegovina, which consumes approximately half of its electricity from hydropower, are prime examples of the region's exploitation of water resources and dependency on a single renewable resource. In order to attain a more balanced portfolio of renewable energy and energy efficiency technologies, economic incentives are needed to stimulate the market for new technologies through investments.

Table 3 provides a summary of current renewable energy consumption, projects, and support mechanisms within selected SEE countries. The most common economic incentives are feed-in-tariffs, which provide a fixed price for electricity generated from renewable energy sources sold to the network or grid. Each feed-in-tariff is technology specific and the use of these preferential tariffs has been shown to help in delivering installed capacity throughout Europe, particularly for wind energy in Germany and Denmark. Feed-in-tariffs have been developed within most of the SEE countries with the exceptions of Bosnia and Herzegovina, Montenegro, and Romania. The technologies specified may range from including all renewable energy sources, such as implemented within Croatia and the Former Yugoslav Republic of Macedonia (Albania is also in the process of developing a feed-in-tariff system for all renewable energy sources), to including only one or two technologies, such as currently in Albania (small hydropower plants) and UNMIK (small hydropower plants and wind).

In many cases, in addition to feed-in-tariffs, national regulatory agencies issue green certificates, which serve as a "guarantee of origin" for renewable energy production. The green certificates are less developed than feed-in-tariff regulations and they are implemented in Bulgaria, Romania, and the Former Yugoslav Republic of Macedonia. Green certificate systems are currently being developed in Albania, Croatia, and Montenegro. Additionally, energy obligations are used in Albania, Croatia, and the Former Yugoslav Republic of Macedonia in order to require the energy supplier to supply a certain percentage or amount of their total supply from approved renewable sources. Finally, several regulatory grid system features, such as connection priority, costs, and usage as well as grid and market expansion, are incorporated throughout SEE countries; however, Montenegro and the UNMIK are still considering these grid system issues.

Government resources towards an environment or energy efficiency fund for renewable and energy efficiency projects have been provided in Albania, Bulgaria, Croatia and Romania. Specifically, the Government of Croatia has implemented a Project Development Fund and Partial Financial Guarantees Fund aimed at providing resources for the creation of demand and a sustainable market for energy efficiency products and services (Stritih et al. 2007). In addition, the European Investment Bank (EIB) Kreditanstalt für Wiederaufbau (KfW), the European Commission, and the European Bank for Reconstruction and Development (EBRD) have recently created a 95 million Euro fund, the Southeast Europe Energy Efficiency Fund (SE4F), that will invest in small- and medium-sized business and household development of renewable energy and energy efficiency projects in Albania, Bosnia, Croatia, Macedonia, Montenegro, Serbia and Turkey (BalkanInsight 2009).

However, despite the important progress of SEE countries in providing support mechanisms for renewable energy and energy efficiency development, the markets for these technologies remain at an early stage of development (Jelavic and Segon 2008).

Table 3 Renewable energy consumption, projects, and renewable energy and energy efficiency support mechanisms

Country	Renewable energy consumption and projects	Feed-in-tariff	Guarantee of origin/green certificates	Grid system issues	Relevant energy fund
Albania[a]	Mostly hydropower consumption (98%) with small HPP, wind, and biofuels projects proposed[b]	Small HPP, All RES in Draft Law (as of Oct 2009)	Both in Draft Law (as of Oct 2009)	Connection priority, Public Supplier obligation to purchase electricity from all eligible RES, Investor covers grid costs	Energy Efficiency and Renewables Fund, Investments from SE4F (Dec 2009)
Bosnia and Herzegovina[c]	Mostly hydropower consumption (~50%) with small HPP, wind, and biofuels projects proposed	Draft Decision, including co-generation (as of Oct 2009)	No	Connection Rules (Electric facilities pay 50% connection fee)	–
Bulgaria[d]	Small HPP, wind, PV, and geothermal projects proposed	Yes (Solar, PV, Wind)	Yes	Connection, Usage, and Grid Expansion Regulations	Energy Efficiency Fund
Croatia[e]	Mostly hydropower; Biomass, solar, wind, 69 RES projects & 218 energy efficiency (2004–9)	Yes	Partly, under development	Connection Rules, Public Supplier obligation to purchase electricity from all eligible RES	Energy Efficiency Fund, Investments from SE4F (Dec 2009)
Montenegro[f]	Mostly hydropower; small HPP and wind projects proposed	Draft Bylaw (as of October 2009)	Draft Bylaw (as of October 2009)	Draft: Priority connectio for RES, Access to grid on non-discriminatory basis, Investor covers grid costs	Investments from SE4F (Dec 2009)
Romania[g]	Mostly small HPP and wind, some biomass projects proposed	No	Yes (quota system[h])	Connection, Usage, and Grid Expansion Regulations	Energy Efficiency and Environment Fund

(continued)

Table 3 (continued)

Country	Renewable energy consumption and projects	Feed-in-tariff	Guarantee of origin/green certificates	Grid system issues	Relevant energy fund
Serbia[i]	Mostly small HPP and wind projects proposed	Yes (end of 2009)	No	Market rules and grid rules under development, No priorities for RES installations	Investments from SE4F (Dec2009)
The Former Yugoslav Republic of Macedonia[j]	Mostly small HPP with some biodiesel and wind projects proposed	Yes (small HPP, wind, PV)	Yes (Rulebook)	Public Supplier obligation to purchase electricity from all eligible RES	Investments from SE4F (Dec 2009)
UNMIK	–	Yes (small HPP and wind)	No	Measure to guarantee transmission and distribution of RES to be introduced	–

Adapted from Uzanov 2008

HPP Hydropower Plant (installed capacity less than 15 MW), *SE4F* Southeast Europe Energy Efficiency Fund, *PV* Photovoltaic

[a] Energy Community, Recent Development on the Alabanian RES Policies, Presentation to the 1st Meeting of the RETF, October 2009
[b] EBRD, Bosnia/Herzegovina Country Profiles (2009)
[c] Energy Community, Bosnia and Herzegovina - Renewable Energy Policy, Presentation to the 1st Meeting of the RETF (2009e)
[d] Bundesministerium für Umwelt, Naturschutz und Reaktorsicherheit, Legal Source on Renewable Energy: Promotion in Bulgaria (2009a)
[e] Energy Community, Overview of Legislative and Institutional Framework in the Republic of Croatia Concerning Renewable Energy Sources, Presentation to the 1st Meeting of the RETF, October 2009
[f] Energy Community, Renewable Energy Policy in Montenegro, Presentation to the 1st Meeting of the RETF, October 2009
[g] Bundesministerium für Umwelt, Naturschutz und Reaktorsicherheit, Legal Source on Renewable Energy: Promotion in Romania (2009c)
[h] Suppliers submit green certificates as evidence of obligations in quota system
[i] Energy Community, Implementation of EU Acqui on Renewables, Presentation to the 1st Meeting of the RETF, October 2009
[j] Energy Community, Renewable Energy Task Force 1st Meeting, Presentation to the 1st Meeting of the RETF, October 2009

Several forms of technology transfer through inward investments and capacity building (which is further discussed in Sect. 4) provide significant opportunities to overcome the barriers to the uptake of renewable energy and energy efficiency technologies and market development.

4 Technology Transfer and Capacity Building

Technology transfer and capacity building provide valuable opportunities for SEE countries to learn from experiences of other countries that have effectively implemented renewable energy and energy efficiency policies, programmes, and projects. It is possible to leverage existing organizations and mechanisms for technology transfer and capacity building, such as (1) the Energy Community Renewable Energy Task Force (RETF), (2) the Southeast Europe Transnational Cooperation Programme, and (3) non-governmental organization (NGO) networks. Additionally, inward investments in renewable energy and energy efficiency projects reveal significant opportunities for technology and knowledge transfer and capacity building. There is a high potential for financing relevant energy projects through the UNFCCC Clean Development Mechanism (CDM) and driving inward investment by creating a favourable environment for business.

The Energy Community RETF Work Programme for 2009–2011 indicates several initiatives that will build capacity among Contracting Parties. The RETF proposes to provide training sessions for ministry officials involved in the renewable energy area as well as renewable energy, energy, and regulatory agencies and authorities (Energy Community 2009d). These training sessions provide an opportunity for SEE government stakeholders to understand the implementation of Directive 2009/28/EC and identify target areas or issues where the advance of knowledge is necessary for all the Contracting Parties. A similar role for capacity building could be incorporated within the Energy Community EETF as Contracting Parties begin plans for the transposition of the recent Ministerial Council Decision to incorporate the Directives relevant to energy efficiency within individual countries' legislation.

The Southeast Europe Transnational Cooperation Programme is a network specific to the SEE region and includes Albania, Austria, Bosnia and Herzegovina, Bulgaria, Romania, Croatia, Macedonia, Greece, Hungary, Serbia, Montenegro, Slovakia, Slovenia, and Moldova (SEETCP 2009a). The Programme is focused on the development of transnational cooperation for various priorities including the environment and the promotion of resource and energy efficiency. Currently, the Programme supports environmental projects as the majority of the total approved projects. However, within these projects, only four renewable energy or energy efficiency related projects and partnerships have been formed (SEETCP 2009b). Since the Programme provides opportunities, tools, resources, and a forum to promote partnerships in common development interests, it may prove an effective venue for the increased uptake of transnational action and synergies in renewable energy and energy efficiency project development.

Leveraging the knowledge of NGOs presents another opportunity for capacity building. For example, the Renewable Energy and Energy Efficiency Partnership (REEEP) is a global coalition of governments, business, and organizations, in addition to working with many SEE governments (REEEP 2009). REEEP assists governments in creating favourable regulatory policies, provides innovative renewable and energy efficiency financing, and supports policy-maker and government networks, such as the Energy Efficiency Coalition (EEC), the Sustainable Energy Regulation Network (SERN) and Renewable Energy and International Law (REIL). Moreover, in 2004, the Regional Environmental Center for Central and Eastern Europe (REC) became the regional secretariat of the REEEP for Central and Eastern Europe and Turkey and currently has 17 country offices throughout the Central and Eastern European region (REC 2009a). The REC is an international organization that aims at solving environmental problems by supporting cooperation among all stakeholders in the environmental decision-making process. Several programmes intersect with capacity building goals in areas relevant to clean energy, such as eco-innovation (Promoting Financing Mechanisms for Eco-innovation in SEE funded from May 2008 to April 2009) and climate change (Preparation of the SEE/CCFAP – Climate Change Framework Action Plan for the SEE Region funded from January to December 2008) (REC 2009b). Therefore, programmes focused on cooperation in sharing development strategies for renewable energy and energy efficiency and compliance with relevant international energy and environmental policies could be developed under the REC.

Furthermore, Agree.Net is a network of non-governmental organizations within the Central, Eastern, and SEE regions that works for the promotion of sustainable energy through the development of projects, actions, and campaigns at the local, national, and international levels (Agree.Net 2007). Although the majority of Agree.Net's efforts are focused on raising public awareness, which addresses the social dimension of a renewable energy and energy efficiency technology market, some projects emphasize the transfer of knowledge and aim at establishing institutional infrastructures that aid in policy, technology, and market development. An example of this is the development of energy efficiency advisory centres in Bosnia and Herzegovina, Hungary, and the Former Yugoslav Republic of Macedonia aimed at transferring knowledge among other local or national NGOs and identifying potential projects and cooperation (Agree.Net 2007). There may be opportunities for an increased role of Agree.Net, or similar NGOs, in leveraging the knowledge base of regional NGOs, while also aiding in related energy policy decisions and garnering support for their implementation.

Financing of renewable energy and energy efficient projects is a key aspect for building an effective and efficient market and regime. With limited government capital allocations and public funding in SEE for environmental projects, generating energy financing is a necessity (Stritih et al. 2007, p. 80). Additionally, even where there are national intentions to contribute more capital to the energy sector, SEE countries' actual ability to contribute may be restricted by overall GDP, lack of economic stability, and budgetary constraints (Stritih et al. 2007).

Significant opportunities exist in attaining relevant energy financing through inward investments in technology transfer and capacity building through the

UNFCCC CDM. The CDM provides an opportunity for candidate host countries within the SEE region to develop approved energy projects. The application of CDM throughout the SEE region has so far been limited, with the majority of registered projects being developed in Latin America and Asia (UNFCCC 2010). However, there is currently one CDM project being developed in the Former Yugoslav Republic of Macedonia, which involves the Netherlands – the Skopje Cogeneration Project (UNFCCC 2009a). This project proposal is undergoing corrections following a recent UNFCCC review. Additionally, Albania has established a bilateral agreement with Italy, and the Former Republic of Macedonia has signed agreements with Italy and Slovenia for the development of CDM projects (Kodzoman 2008). Albania has also been receiving assistance from Austrian Development Assistance in building Albania's capacity to access carbon finance through regulatory procedures. Austrian Development Assistance is aiding Albania in developing two CDM Program Design Documents for small hydropower plant projects with Austria. The Former Republic of Macedonia has produced a National Strategy for CDM identifying high priority areas for CDM financing through the rehabilitation of large power plants, fuel switching to natural gas, combined heat and power for district heating, industrial efficiency improvements, hydropower, and geothermal energy (UNDP 2007). Moreover, the United Nations Development Programme (UNDP) has provided CDM assistance through capacity building projects and has adopted a MoU with the Former Republic of Macedonia for the development of a Biogas Power Plant with the UNDP Millennium Development Goals Carbon Fund (Kodzoman 2008). Thus, the CDM presents a significant opportunity that should be considered not only within the context of supporting renewable energy and energy efficiency within the SEE region but also as a way to generally promote inward investments in new technologies in the region.

Most often, the implementation of relevant energy policies and projects throughout the SEE region are initiated in cooperation and funding assistance from key multilateral and bilateral donors, such as the EBRD, EIB, UNDP, World Bank, KfW, German Federal Ministry for Cooperation and Development (GTZ), United States Agency for International Development (USAID), and the Greece Ministry of Foreign Affairs' HellenicAID. These donors help propose new or existing policy or institutional mechanisms for the development of renewable energy and energy efficiency projects. For instance, EBRD supports institutional capacity building throughout the SEE region through its Sustainable Energy Initiative (SEI), which works to identify areas for improving regulatory and incentive structures, helps prioritize reforms with country Ministries, Regulators, and transmission or distribution system operators in the electricity sector, and guides entrepreneurs seeking to develop projects (Maly 2009). The USAID also supports training programmes in energy efficiency for local level administrators and institutions. For example, USAID collaborated with the Association of Albanian Municipalities to fund and organize the First Energy and Water Efficiency Training Program for Albanian Municipalities (Leno 2007). Additionally, the USAID coordinates with other donors, such as HellenicAID, in the SYNENERGY Programme to provide institutional network development and address barriers to investment and financing for Energy Community countries (Birnbaum 2009). Through the SYNENERGY Programme, co-financing pilot schemes for solar thermal are currently being planned in Croatia, Bosnia and Herzegovina, and Albania (CRES 2009). Thus, an increased coordination of multilateral and bilateral donor funding and activities could help leverage capacity building resources and knowledge provided at the local or national levels.

Reforming and refining the national regulatory and incentive structures are important factors to overcome barriers to investments and drive foreign direct investment (FDI) in renewable energy and energy efficiency projects. A World Bank (2009) report identifies several criteria and measures that can be implemented in order to generally improve the environment for business development. Criteria for business development that are analyzed within the report include quantitative measures of regulations for starting a business, dealing with construction permits, employing workers, registering property, getting credit, protecting investors, paying taxes, trading across borders, enforcing contracts and closing a business (World Bank 2009). The Former Republic of Macedonia's ranking in 2010 jumped to 32 from 69 in 2009, and it was ranked as one of the most improved countries in terms of reforming seven criteria including regulations for starting a business, providing credit, and protecting investors, among others (World Bank 2009). Governments may find it worthwhile to focus on business development areas mentioned in the report. This could provide benefits in attracting FDI if regulatory frameworks and economic incentives in energy, as well as other sectors, are well placed.

Furthermore, an example of reforming policy in order to overcome investment barriers in the energy sector is Bosnia and Herzegovina's establishment of the Law on the Policy of Foreign Investment and the Foreign Investment Promotion Agency (FIPA) in 1998 (Paoletti 2009). FIPA aims at attracting and maximizing the flow of FDI and facilitating the interaction between public and private sectors in order to improve the investment environment and promote economic and regulatory stability for investors. This experience provides a model for other SEE countries. However, in its current state, FIPA resources acquired through its Foreign Investor Support Fund (FISF) do not support energy sector projects (Paoletti 2009). In order to prioritize FDI in the clean energy sector, funding requirements providing sufficient resources to support energy projects should be enacted alongside the adoption of FDI policies. Nonetheless, this model provides a valuable experience of FDI policy reform within the region with lessons to be learned in the context of developing relevant energy projects if implemented within other SEE countries.

5 Institutional Development and Cooperation

5.1 National and Local Institutions

Institutional capacity and structures are probably the most important aspect of an effective renewable energy and energy efficiency regime (Stritih et al. 2007, p. 57). Thus, building strong, well-defined, national and local institutions will establish a key pillar to developing an effective regime. To clarify, in speaking of institutional capacity we are referring to national SEE institutions' ability to develop and implement renewable energy or energy efficiency initiatives. To date, most SEE countries have at least established national ministries with jurisdiction over energy issues. These ministries include, for example, (1) the Albanian Ministry of Industry and Energy, (2) the Austrian Federal Ministry of Economic Affairs and Labour, (3) the Bosnia and Herzegovina Ministry of Foreign Trade and Economic Relations, (4) the Croatian Ministry of Economy, Labour and Entrepreneurship, (5) the Bulgarian Ministry of Energy and Energy Resources, (6) the Macedonian Ministry of Economy Department of Energy and Mineral Resources, (7) the Montenegro Ministry of Economy, (8) the Romanian Ministry of Economy and Trade, (9) the Serbian Ministry of Mining and Energy, and (10) the Slovenian Ministry of Environment, Spacial Planning, and Energy (SEEnergy 2009).

More importantly, however, several SEE countries have established agencies specifically geared to manage energy savings: (1) the Energy Efficiency Centre (EEC) of Albania; (2) the Austrian Energy Agency (AEA); (3) the Energy Efficiency Agency of Bulgaria; (4) Energy Institute 'Hrvoje Pozar' of Croatia; (5) the Romania Agency for Energy Conservation; (5) the Serbian Energy Efficiency Agency; and (6) the Agency of Republic of Slovenia for Energy Efficiency and Renewable Energy Sources (AURE) (Stritih et al. 2007). In addition, several SEE countries have also established regulators that, among other things, create secondary legislation, set energy prices, issue and revoke licenses, and handle dispute settlements involving national energy market issues (Stritih et al. 2007). Therefore, in many ways, the basic institutional foundation for successful adaptation of the energy sector is currently in place in many SEE countries.

Nonetheless, despite their development, most SEE energy institutions face challenges resulting primarily from their lack of cohesion and limited resources. First, most SEE countries have more than one environmental institution and coordination among various institutions is limited. This can make policy co-ordination and the implementation of efficient new legislation difficult. To be sure, because it is difficult to identify clear responsibility for environmental issues, a more integrated approach to addressing environmental concerns is required to ensure adequate institutional accountability (Stritih et al. 2007, pp. 63–64). Inherently, any new policies will be diluted and less efficient and/or effective without establishing further transparency and accountability mechanisms. Second, in terms of resources, SEE national finance ministries are often reluctant to grant funds for new projects because of limited government resources (Stritih et al. 2007, p. 63). Compared with other social issues, energy efficiency and other environmental issues tend to receive less political attention from national governments (Stritih et al. 2007, p. 64). Further, in terms of strengthening and maintaining strong SEE energy efficiency institutions, difficulties exist because low public sector salaries create problems attracting necessary talent (Stritih et al. 2007, p. 64). Thus, because the private sector is inherently more lucrative, talented individuals that could make important contributions to the energy efficiency movement at the government level have little incentive to do so with the SEE region. Third, effective implementation and monitoring of new energy policies strategies is difficult without people on the ground to carry out and enforce those policies and strategies. Therefore, with limited national funding to funnel down to local authorities, implementation and monitoring can be difficult. In sum, because these challenges have yet to be overcome by SEE energy institutions, opportunities exist for further development of institutions.

In part, stronger local and sub-national institutions could help address the institutional challenges described (Stritih et al. 2007, pp. 66–72). Not only can these types of institutions help implement and carryout the policies and strategies of national institutions on the ground, they can also create both local-level competition and cooperation, which can spur the development of effective new energy strategies and ideas that can make their way into national policies. Competition will result from the desire of local institutions to attract FDI to help develop particular regions. At the same time, cooperation will result from the need to address shared energy concerns and manage transboundary environmental resources. In this way, the resulting mix of local competition and cooperation can help effectively adapt to future energy challenges in the SEE region by encouraging the development of innovative strategies. Moreover, in terms of addressing limited financing within the region, local institutions can help support national energy efforts by working together to attract FDI from outside the SEE region to help building local and national capacity. Ultimately, building stronger local institutions can complement existing national institutions and help national institutions to address some of the challenges discussed. Coordination and guidance could come from the national institutions, while sub-national institutions could provide manpower, on-the-ground understanding of local energy challenges, and new ideas to confront those challenges.

Still, because some of the challenges faced by SEE institutions simply cannot, and will not, be corrected through internal measures (e.g. the strengthening of sub-national institutions and increased accountability at the national level), other measures must support or incentivize national and local government efforts. Therefore, regional cooperation among SEE national and sub-national institutions and international support are necessary if SEE countries are going to successfully adapt to confront future energy challenges.

5.2 *Regional Cooperation and International Support*

Despite the great differences existing among SEE countries (e.g. differing levels of economic development, different political models, etc.), SEE countries do face similar energy security concerns, which can establish an effective baseline for cooperation (Ozunu et al. 2007). Because environmental issues can, and do, present problems for multiple countries, there is an inherent need for collective action. The environmental problems of one country regularly become the environmental problems of surrounding countries (Mihajilov 2008). This geopolitical reality is especially true in the area of energy security.

To be sure, the SEE region's dependence on Russian gas supplies creates a shared need to promote and develop alternative forms of energy (Crooks 2009). As Russia's winter 2009 reduction of gas deliveries to Europe suggested, the SEE region does share similar energy concerns (Synovitz 2009). When these deliveries were reduced, several SEE countries were negatively affected. For instance, Slovakia "lost 97 percent of its supplies," while Serbia and Bosnia-Herzegovina were also "hit hard by the gas cuts – subjecting industrial users to severe rationing and forcing many people to live without heat" (Synovitz 2009). Further, because "Bulgaria receives 92 percent of its natural gas from Russia via pipelines that pass through Ukraine," it was also "hit hard by halted gas deliveries" (Synovitz 2009). Consequently, regardless of who is to blame for this gas reduction incident, the outcomes showed that these and other SEE countries should be collectively considering their energy alternatives.

At the same time, the end goal of many SEE countries is EU accession. Yet, EU membership requires, among other things, compliance with the Energy Community Treaty and implementation of the *aquis communautaire*. As a result, there is a collective need to coordinate energy-related policies and programmes. And thus, in the course of pursuing EU membership, it is practical for SEE countries to collaborate to effectively implement the *aquis communautaire*.

Moreover, how can regional cooperation help correct the national institutional challenges we previously discussed (e.g. lack of coordination, accountability, financing, and enforcement tools)? Initially, through cooperation, SEE countries establish that shared energy challenges exist, and national institutions recognize the need for flexibility in adapting the energy sector. In particular, national institutions are suggesting they are willing to consider various regulatory and economic solutions to energy problems, which is necessary for effective energy-sector adaptation.

More importantly, however, continued and deeper regional cooperation can help establish greater accountability, transparency, rule of law, and better enforcement mechanisms, which encourage more FDI. Specifically, these types of national challenges are exactly what regional institutions seek to address. For instance, the Energy Community Treaty specifically aims at creating a stable regulatory and market framework that (1) attracts investment in power generation and networks in order to ensure stable and continuous energy supply that is essential for economic development and social stability, (2) creates an integrated energy market allowing cross-border energy trade and integration with the EU market, (3) enhances the security of supply, (4) improves the environmental situation in relation with energy supply in the region, and (5) enhances competition at regional level and exploits economies of scale (Energy Community Treaty 2005, Article 2). Thus, the Treaty seeks to promote greater transparency, accountability, and rule of law through regional cooperation that is intended to attract more FDI into the SEE region, particularly in the energy sector.

In addition, the Treaty provides a dispute settlement mechanism, which further enhances national transparency and accountability in the energy sector by establishing rules for tackling a breach of the Treaty by a member country (Energy Community Treaty 2005, Article 89–93). In other words, the Treaty seeks to increase transparency by exposing national policies and to increase accountability by requiring national institutions to answer to other member countries for their decisions. But despite these cooperative strides and the substantial benefits derived from the Treaty, it should be noted that both continued and new collaborative efforts will be needed to confront future SEE energy challenges. In particular, because "nothing in th[e] Treaty shall affect the rights of a Party to determine the conditions for exploiting its energy resources, its choice between different energy sources and the general structure of its energy supply," there is room for manoeuvrability by national institutions, which means continued efforts to improve transparency, accountability and the rule of law are necessary (Energy Community Treaty 2005, Article 8). In addition, further cooperative efforts will be required to ensure the appropriate local enforcement mechanisms are in place.

To summarize, in an already socially and politically fragile SEE region, a lack of regional energy-sector cooperation will likely result in an increased struggle for energy security, and this "could serve as a catalyst for resorting to violence and becoming a risk for state (and regional) security and stability" (Mihajilov 2008). Because SEE countries individually lack the ability to confront SEE energy security concerns (for the reasons previously discussed), regional cooperation is essential. Regional cooperation can help overcome some of the challenges faced by national institutions. And because many energy-related problems across the SEE region are similar, the baseline for cooperation is, for the most part, already established: there is a foundation for negotiation in this area. Finally, the benefits derived from working together (e.g. ability to overcome national impediments to energy-sector adaptation and to address shared environmental challenges in pursuit of EU membership) help incentivize collective action. If the SEE region is going to effectively confront future energy-sector adaptation challenges, regional cooperation needs to continue and build on the efforts already underway.

Finally, in addition to regional cooperation and the institutional support being provided by the EU, which was already discussed in the context of the Energy Community Treaty, international support from both inside and outside the EU can help overcome national SEE energy-sector challenges. While the primary form of international support is typically thought to be international financing, constructive support is not limited to financial contributions. Both public and private international players can provide novel ideas for adapting the SEE energy sector. For example, knowledge sharing by governments and international organizations with SEE countries can help SEE countries avoiding the mistakes of others. It can also help eliminating the inefficient use of valuable, and limited, resources on efforts that have been unsuccessful in other countries. In this way, resources can be directed towards programmes that are likely to have the greatest impact given the investment. Finally, international support can create new opportunities through physical technology sharing and capacity building, as previously discussed. Consequently, along with regional cooperation, international support has a valuable role to play in contributing to SEE energy adaptation efforts.

6 Conclusion

This paper has briefly introduced the relevant energy policies and economic incentives for renewable energy and energy efficiency within the SEE region. In doing so, it has compared the current SEE energy regime with five fundamental components of an effective and workable energy regime. In particular, it has focused on opportunities for technology transfer and capacity building, national institutions, and regional cooperation. Based on our assessment of the current regime, countries in the SEE region have made significant progress towards implementation of the *aquis communautaire* and the development of renewable energy and energy efficiency policies, programmes, and projects. Nevertheless, there is still opportunity for improvement. Primarily, opportunities exist where the most difficult challenges or barriers to effective SEE energy policy remain. These barriers include, among other things, lack of stable legislation, support mechanisms to attract FDI, coordination by national agencies responsible for renewable energy and energy efficiency policies, and effective cooperation among existing national and regional institutions.

Accordingly, to promote a stronger renewable energy and energy efficiency regime, it is important to seize opportunities for technology transfer and capacity building within the existing regional networks, such as through the Energy Community Renewable Energy Task Force (RETF) and Energy Efficiency Task Force, the Southeast Europe Transnational Cooperation Programme, and non-governmental organizations that are already working in a trans-national context. In the area of CDM, there is a high potential to drive inward investments and innovative financing schemes for renewable energy and energy efficiency projects. A multitude of opportunities exist in this area given the present development of the clean energy sector throughout the EU and the growing awareness of clean energy benefits in helping achieve possible emission targets after EU accession. SEE countries could work to create a favourable financing environment to advance the development and learning of new technologies. For example, governments may find it worthwhile to focus on business development areas mentioned in the World Bank (2009) report, in which the Former Republic of Macedonia was ranked as one of the most improved countries in terms of starting business, providing credit, and protecting investors, among other criteria. Improvement in these areas could provide benefits in attracting FDI if economic incentives in this sector are well placed.

Furthermore, based on our review of the SEE energy-sector institutions and collaborations, there is an abundance of ongoing endeavours in the clean energy and energy conservation sector. In the SEE region, national, regional, and international institutions will need to further collaborate and build on the identifiable policies and programmes that are currently underway. Yet, it seems that many of these endeavours are not coordinated with one another. Therefore, there is a greater need for coordination among these ongoing projects created by national, regional, and international policies and organizations. There is a need for more coordination within the existing network of national and international energy and environmental policies and obligations. This may extend beyond mere collaboration between institutions and could incorporate how national-level best-practices and initiatives can be scaled-up or build on lessons learned at the regional or international level. In this way, greater resources and knowledge may be leveraged. Additionally, coordination of national or local-level renewable energy and energy efficiency initiatives can result in more geographical opportunities for technology market expansion and a greater knowledge pool for technology development.

Moreover, this research has prompted us to develop a list of additional questions that we were unable to contemplate in this brief paper but that we think deserve further attention. In the context of economic incentives, it would be interesting to explore what combination of support mechanisms, and at what stage of market development, would be needed for the advancement of specific technologies in the SEE region. In addition, establishing a working list of existing and needed 'best practices' to promote uptake of renewable energy and energy efficiency technology that is appropriate (in terms of resource potential, legislative and institutional capacity) for individual SEE countries could provide an added focus on how the SEE region could move forward in this area. In terms of technology transfer and capacity building, what are the best methods for building capacity and attracting FDI in renewable energy and energy efficiency technologies? In the context of institutions, if established, what role might a Regional Energy Agency play in strengthening current energy development plans within the SEE region? And more generally, what framework should be used to determine whether a SEE energy agency is effective – should there be flexibility in devising such a framework and would it look different than a framework used to measure the effectiveness of an energy agency in a more developed region? Which are the effective methods of leveraging energy, environmental, climate change, and national security programmes and coordinating among the various existent national and international networks? Finally, when we talk about regional cooperation, to what extent is bilateral cooperation more, or less, effective than multilateral collaborations (e.g. should EU countries be part of this process)? And, are there other external conditions that are necessary for successful cooperation within the SEE region?

While addressing these questions is beyond the scope of this paper, the questions are worth considering in the context of SEE renewable energy and energy efficiency policy. If the SEE region is going to effectively adapt its energy sector to ensure future energy security, these questions, and many others, will have to be considered so that policy makers can optimize decisions and leverage existing resources, institutions, and networks within the region.

References

Albania Ministry of Industry and Energy, National Agency of Energy (NAE) (2005) Summary – the national strategy of energy and plan of action. UNFCCC. http://unfccc.int/files/meetings/seminar/application/pdf/sem_albania_sup1.pdf. Accessed 25 Dec 2009

Agree.Net (2007) Annual report 2007: promoting sustainable energy to address climate change. Agree.net, Czech Republic

American Chamber of Commerce Kosovo (ACCK) (2008) Implementation plan of the Kosovo energy strategy for the period 2009–2011. Presentation to the Rochester institute of technology/American University in Kosovo Center for Energy and Natural Resource Development, 22 December

BalkanInsight (22 Dec 2009) New Fund Aims to Cut Balkan Carbon Footprint. http://www.balkaninsight.com/en/main/news/24578. Accessed 27 Dec 2009

Birnbaum I (2009) USAID assistance on energy efficiency and renewable energy under SYNENERGY Cooperation with HellenicAID. Presentation at the renewable energy task force 1st meeting, 8 October. http://www.energy-efficiency.me/presentations/thursday/theme_3/ira_birnbaum.pdf. Accessed 22 Oct 2010

Bundesministerium für Umwelt, Naturschutz und Reaktorsicherheit (2007) Legal source on renewable energy: promotion in Austria. http://res-legal.eu/en/search-for-countries/austria/more-about/land/oesterreich/ueberblick/foerderung.html. Accessed 21 Dec 2009

Bundesministerium für Umwelt, Naturschutz und Reaktorsicherheit (2009a) Legal source on renewable energy: promotion in Bulgaria. http://res-legal.eu/en/search-for-countries/bulgaria/more-about/land/bulgarien/ueberblick/foerderung.html. Accessed 21 Dec 2009

Bundesministerium für Umwelt, Naturschutz und Reaktorsicherheit (2009b) Legal source on renewable energy: promotion in Hungary. http://res-legal.eu/en/search-for-countries/hungary/more-about/land/ungarn/ueberblick/foerderung.html. Accessed 21 Dec 2009

Bundesministerium für Umwelt, Naturschutz und Reaktorsicherheit (2009c) Legal source on renewable energy: promotion in Romania. http://res-legal.eu/en/search-for-countries/hungary/more-about/land/ungarn/ueberblick/foerderung.html. Accessed 21 Dec 2009

Center for Renewable Energy Sources (CRES) (2009) Hellenic aid activities for RES promotion in the energy community: the common USAID-HellenicAID programme SYNENERGY. Presentation at the renewable energy task force 1st meeting, 8 October. http://www.energy-community.org/pls/portal/docs/428183.PDF. Accessed 22 Oct 2010

Crooks E (ed) (2009) Understanding energy policy: problem to combat. Financial Times, 2: 4

Directive 2009/28/EC of the European Parliament and of the Council (23 April 2009) http://eur-lex.europa.eu/LexUriServ/LexUriServ.do?uri=OJ:L:2009:140:0016:0062:EN:PDF. Accessed 20 Dec 2009

Energy Community (2009a) Annual report on the activities of the energy community to the parliaments. http://www.energy-community.org/pls/portal/docs/442180.PDF. Accessed 29 Dec 2009

Energy Community (2009b) Decision No. 2009/02/MC-EnC of 18 December 2009 on the implementation of certain directives on energy efficiency. http://www.energy-community.org/pls/portal/docs/474205.PDF. Accessed 20 Dec 2009

Energy Community (2009c) Facts and figures. http://www.energy-community.org/pls/portal/docs/278177.PDF. Accessed 29 Dec 2009

Energy Community (2009d) Proposed activities related to renewable energy: a work programme outline 2009–2011. http://www.energy-community.org/pls/portal/docs/428185.PDF. Accessed 7 Jan 2010

Energy Community (2009e) Bosnia and Herzegovina – Renewable energy policy. Presentation to the 1st meeting of the RETF, energy community, 8 October. http://www.energy-community.org/pls/portal/docs/428190.PDF. Accessed 22 Oct 2010

Energy Community (2009f) Report of the renewable energy task force chair on the activities related to renewable energy. http://www.energy-community.org/pls/portal/docs/474199.PDF. Accessed 20 Dec 2009

Energy Community (2009g) Report to the 6th ministerial council on work performed by the energy efficiency task force. http://www.energy-community.org/pls/portal/docs/356185.PDF. Accessed 20 Dec 2009

Energy Community (2009h) Report to the 7th ministerial council on the energy efficiency task force. http://www.energy-community.org/pls/portal/docs/474200.PDF. Accessed 20 Dec 2009

European Bank for Reconstruction and Development (EBRD) (2009a) Bosnia/Herzegovina country profiles. http://ebrdrenewables.com/sites/renew/countries/BosniaHerzegovina/default.aspx. Accessed 29 Dec 2009

European Bank for Reconstruction and Development (EBRD) (2009b) Bulgaria country profiles. http://ebrdrenewables.com/sites/renew/countries/Bulgaria/default.aspx#projects. Accessed 29 Dec 2009

European Bank for Reconstruction and Development (EBRD) (2009c) Montenegro country profiles. http://www.ebrdrenewables.com/sites/renew/countries/Montenegro/default.aspx. Accessed 29 Dec 2009

Jelavic B, Segon V (2008) Report on the implementation of the acquis on renewables in the energy community contracting parties: final report presentation. Presentation to the promoting renewable energy in South Eastern Europe through international cooperation, 18 March. http://archive.rec.org/reeep/docs/meetings/2008_march_17_18/presentations/branka_jelavic.ppt. Accessed 22 Oct 2010

Karafilovska M (2009) Renewable energy task force 1st meeting. Presentation to the 1st meeting of the RETF, Energy Community, 8 October

Kodzoman A (2008) New potentials for renewable energy projects in the Balkans through CDM. Presentation at the workshop on promoting renewable energy in South Eastern Europe through international cooperation, 18 March. http://archive.rec.org/reeep/docs/meetings/2008_march_17_18/presentations/anita_kodzoman.ppt. Accessed 22 Oct 2010

Kovacevic I (2009) Renewable energy policy in Montenegro. Presentation to the 1st meeting of the RETF, Energy Community, 8 October

Leno V (2007) Energy efficiency in Albania. Presentation in Prishtina, 2 November. http://www.rit.edu/research/cenr/documents/Vaso%20Leno%20Energy%20Efficiency.pdf. Accessed 22 Oct 2010

Maly M (2009) EBRD Institutional capacity building for western Balkans. Presentation at the renewable energy task force 1st meeting, 8 October. http://www.energy-community.org/pls/portal/docs/428180.PDF. Accessed 22 Oct 2010

Mihajilov A (2008) Regional environmental initiative: South-Eastern Europe (Balkan) regional environmental cohesion initiative. Presented at Pan-European Conference on EU Politics, Riga, Latvia, 25–27 Sept 2008

Organization for Economic Co-operation and Development (OECD) (2008) An OECD framework for effective and efficient environmental policies: overview. OECD. http://www.oecd.org/dataoecd/8/44/40501159.pdf. Accessed 20 Nov 2009

Ozunu A, Cadar D, Crina Petrescu D (2007) Environmental security: priority issues and challenges in Romania and European transition countries. In: Hull RN et al (eds) Strategies to environmental security in transition countries. Springer, the Netherlands, pp 13–23

Paoletti F (2009) Incentives for foreign investments – case study of Bosnia and Herzegovina. Presentation at the United Nations Economic Commission for Europe, 10–11 November

Raguzin I (2009) Overview of legislative and institutional framework in the Republic of Croatia concerning renewable energy sources. Presentation to the 1st meeting of the RETF, Energy Community, 8 October

Regional Environmental Center for Central and Eastern Europe (REC) (2009a) Directory of projects for All SEE countries. http://archive.rec.org/REC/Introduction/intro.html. Accessed 8 Jan 2010

Regional Environmental Center for Central and Eastern Europe (REC) (2009b) Regional REEEP secretariat for central and Eastern Europe and Turkey. http://archive.rec.org/reeep. Accessed 7 Jan 2010

Renewable Energy and Energy Efficiency Partnership (REEEP) (2009) Welcome to REEEP. http://www.reeep.org. Accessed 7 Jan 2010

SEEnergy (2009) http://www.seenergy.org/index.php?/countries/all_countries&type=3. Accessed 17 Dec 2009

Southeast Europe Transnational Cooperation Programme (SEETCP) (2009a) Programme area. http://www.southeast-europe.net/en/about_see/participating_countries. Accessed 7 Jan 2010

Southeast Europe Transnational Cooperation Programme (SEETCP) (2009b) Projects. http://www.southeast-europe.net/en/projects/approved_projects/?call_no=1&priority_2=2&area_of_intervention[]=7&all_regions=1&x=27&y=12. Accessed 7 Jan 2010

Stojadinovic D (2009) Implementation of EU acqui on renewables. Presentation to the 1st meeting of the RETF, energy community, 8 October. http://www.energy-community.org/pls/portal/docs/428191.PDF. Accessed 22 Oct 2010

Stritih J, Qirjo M, Cani E, Myftiu A, Spasojevic D, Stavric V, Markovic M, Simic D, Deda S, UNDP Croatia Environmental Governance Team (2007) Environmental policy in South-Eastern Europe. United Nations Development Programme. http://europeandcis.undp.org/environment/seeurope/show/B341E335-F203-1EE9-BC2D83CD5B55F495. Accessed 22 Oct 2010

Synovitz R (14 January 2009) With gas cut off, Southeast Europe questions reliance on Russia. In Radio free Europe/radio liberty. http://www.rferl.org/content/With_Gas_Cut_Off_Southeast_Europe_Questions_Reliance_On_Russia/1370013.html. Accessed 8 Jan 2010

Treaty Establishing the Energy Community (Energy Community Treaty) (25 Oct 2005). http://www.energy-community.org/pls/portal/docs/530177.PDF. Accessed 22 Oct 2010

Tugu F (2009) Recent development on the Alabanian RES policies. Presentation to the 1st meeting of the RETF, energy community, 8 October. http://www.energy-community.org/pls/portal/docs/428186.PDF. Accessed 22 Oct 2010

United Nations Development Programme (UNDP) (2007) National strategy for clean development mechanism for the first commitment period of the Kyoto Protocol 2008–2012. http://www.undp.org.mk/datacenter/files//files13/nskp.pdf. Accessed 29 Nov 2009

United Nations Environment Programme (UNEP) (2009) From conflict to peacebuilding: the role of natural resources and the environment. UNEP. http://www.unep.org/pdf/pcdmb_policy_01.pdf. Accessed 20 Nov 2009

United Nations Framework Convention on Climate Change (UNFCCC) (2009a) Clean development mechanism project development document: Skopje cogeneration project. http://cdm.unfccc.int/UserManagement/FileStorage/3FHAMY6XKG0T89Z1C4QR52PDWSJNIB. Accessed 5 Jan 2010

United Nations Framework Convention on Climate Change (UNFCCC) (2009b) Kyoto protocol: status of ratification. UNFCCC, 3 December 2009. http://unfccc.int/files/kyoto_protocol/status_of_ratification/application/pdf/kp_ratification_20091203.pdf. Accessed 15 Dec 2009

United Nations Framework Convention on Climate Change (UNFCCC) (2010) Project search. http://cdm.unfccc.int/Projects/projsearch.html. Accessed 5 Jan 2010

Uzanov S (2008) Implementation of the acquis on renewables under the treaty establishing the energy community. Presentation at promoting renewable energy in South Eastern Europe through international cooperation, 18 March

World Bank (2009) Doing Business. Washington, DC: the international bank for reconstruction and development/World Bank.http://www.doingbusiness.org/documents/fullreport/2010/DB10-full-report.pdf. Accessed 3 Dec 2009

Challenges of Environmental Protection in Times of Armed Conflict

Robert Mrljić

Abstract The protection of the natural environment in times of armed conflict is presented through a short analysis of the most relevant concepts and issues typical for this field of international law. Particular attention is placed on the most relevant documents and rules: Art. 35(3) and Art. 55 of Additional Protocol I to the 1949 Geneva Conventions, Environmental Modification Techniques Convention (ENMOD Convention) and Art. 8 (2)(b)(iv) of the Statute of the International Criminal Court (ICC). The relevant practice of the International Court of Justice (ICJ) is also included in the analysis. The individualization of crimes against the environment, the applicability of peacetime treaties in times of armed conflict and the different proposals for enhancing the effectiveness of environmental protection in times of armed conflict are also presented. The protection of water resources and water installations in times of armed conflict is analyzed as well, through the analysis of the 1976 ILA Resolution on the Protection of Water Resources and Water Installations in Times of Armed Conflict and other relevant sources.

Keywords Environmental protection • ENMOD Convention • International Criminal Court • International Court of Justice • Protection of water resources in times of armed conflict

1 Introduction

With the start of the Vietnam War, for the first time public attention was drawn to the issue of environmental protection in times of armed conflict. From this time on, almost every major armed conflict somehow influenced the development of the rules related to the protection of the environment in armed conflict. The purpose of this paper is to offer a brief overview of the existing international legislation in this

R. Mrljić (✉)
University of Zagreb, Zagreb, Croatia
e-mail: rmrljic@pravo.hr

field, as well as to draw attention on some of the most important problems and tendencies in its further evolution. It will start by looking into the most relevant treaties and rules, i.e. Art. 35 (3) and Art. 55 of Protocol I Additional to the 1949 Geneva Conventions (Additional Protocol I),[1] the Environmental Modification Techniques Convention (ENMOD Convention)[2] and Art. 8 (2)(b)(iv) of the Statute of the International Criminal Court (ICC).[3] The scope of protection and specific issues related to each of these documents will be analyzed. This will be followed by a short presentation of other important multilateral treaties and conventions related to environmental protection in times of armed conflict. The most relevant proposals for the enhancement of environmental protection in times of armed conflict will be introduced. As the legal regulations governing this field are in the process of developing their special rules related to specific fields of protection, an analysis of the most relevant rules relating to the protection of water resources and water installations in times of armed conflict will be presented as well.

2 The Additional Protocol I to the Geneva Conventions

The starting point of the following analysis is Art. 35 (3) and Art. 55 of Additional Protocol I.

Art. 35 (3) Basic rules:

> 3. It is prohibited to employ means and methods of warfare which are intended, or may be expected, to cause widespread, long-term and severe damage to the natural environment.

Art. 55 Protection of the Natural Environment:

> 1. Care shall be taken in warfare to protect the natural environment against widespread, long-term and severe damage. This protection includes a prohibition of the use of methods or means of warfare which are intended or may be expected to cause such damage to the natural environment and thereby to prejudice the health and survival of the population.
> 2. Attacks against the natural environment by way of reprisals are prohibited.

The first rule is the result of a slow process of realising that nature itself, i.e. the natural environment has to be protected in case of an armed conflict and not only the human environment. Such understanding of the natural environment, which has to be spared per se from widespread, long-term and severe damage, presents at the same time a turning point in understanding the definition of the environment in accordance with the "intrinsic value approach" of Art. 35, as contrasted with the

---

[1] Protocol Additional to the Geneva Conventions of 12 August 1949, and Relating to the Protection of Victims of International Armed Conflicts (1977) 1125 UNTS.

[2] Convention on the Prohibition of the Use of Environmental Modification Techniques (1977) 1108 UNTS 151.

[3] Rome Statute of the International Criminal Court (1998) 2187 UNTS 3.

"anthropocentric value approach" of Art. 55.[4] The former approach recognizes the environment itself as deserving protection, while the latter links environmental protection to "the health and survival of the population", implying the protection of the human environment. At the same time, Art. 35 (3) departs from the traditional understanding of the "Hague law", which concentrates on the limitation of the means and methods of warfare. On the other hand, Art. 55 departs from the principles of the "Geneva law", which is more oriented towards the protection of the population during an armed conflict. These are also some of the reasons why two articles and not a single one were drafted and finally accepted, although different proposals existed during the *travaux préparatoires*.[5] Indeed, the term "natural environment", as accepted in the wording of these two articles, *"does refer to this system of inextricable interrelations between living organisms and their inanimate environment"*.[6]

In general, there are two groups of different issues related to the application and effectiveness of these two crucial articles. The first one is related to certain doubts and problems which are common for the law of armed conflict in general. The most obvious example is the well-known concept of military necessity and its ambiguity in respect of other principles of humanitarian law and its relation to the protection of the environment.[7] Reprisals prohibited by virtue of Art. 55 (2) belong to the same group of issues common for the law of armed conflict in general.

The second group of issues is associated with the unclarity incorporated in some words used in Art. 35 (3) and Art. 55. First of all, the meaning of "widespread, long-term and severe" is not defined in Additional Protocol I. However, some commentaries defined "long-term" as several decades,[8] while the definition of "widespread" and "severe" is not so clear, although based on the wording used by the *Rapporteur* it can be concluded that the damage caused *"would be likely to prejudice, over a long term, the continued survival of the civilian population or would risk causing major health problems"*.[9] It is important that the standard required for damage in this context is triple and cumulative. This means that the three mentioned aspects of damage, related to each other with the word "and", have to be met in order to reach the threshold of damage as determined by Art. 35 (3) and Art. 55.

The wording "which are intended or may be expected" also leaves possibilities for different interpretations, although it seems clear that it relates to the prohibition of means and methods of warfare which are known to cause damage.[10]

---

[4] See: Schmitt (2000, p. 276).
[5] More in: Sandoz et al. (1987, pp. 411–414).
[6] See: *Id.* (p. 415).
[7] More on military necessity in Additional Protocol I in: *Id.* (pp. 392–396).
[8] See: *Id.* (p. 416).
[9] See: *Id.* (p. 417).
[10] D. Verwey (1995, p. 11).

Although Art. 35 (3) and Art. 55 undoubtedly constitute the basis for environmental protection in times of armed conflict, there are also other relevant provisions of Additional Protocol I. The most relevant provisions are Art. 54 and Art. 56. Art. 54 states the protection of objects indispensable for the survival of the civilian population, such as "foodstuffs, agricultural areas for the production of foodstuffs, crops, livestock, drinking water installations and supplies and irrigation works". On the other hand, Art. 56 is concerned with the protection of works and installations containing dangerous forces in times of armed conflict. The objects of this special protection as set out in Art. 56 are dams, dykes and nuclear electrical generating stations, whereas Art. 35 (2) determines the conditions under which such special protection ceases to exist. Such limited protection of these objects is mostly related to situations in which these objects are used in regular, significant and direct support of military operations and if an attack against these object is the only feasible way to terminate such support.[11] It also has to be mentioned that Article 17 of the Annex I to the Additional Protocol I introduces international special signs for works and installations containing dangerous forces, which is for the first time mentioned in Art. 56 (7).[12]

---

[11] Art. 56 (2) *The special protection against attack provided by paragraph 1 shall cease:*

(a) *for a dam or dyke only if it is used for other than its normal function and in regular, significant and direct support of military operations and if such attack is the only feasible way to terminate such support;*
(b) *for a nuclear electrical generating station only if it provides electric power in regular, significant and direct support of military operations and if such attack is the only feasible way to terminate such support;*
(c) *for other military objectives located at or in the vicinity of these works or installations only if they are used in regular, significant and direct support of military operations and if such attack is the only feasible way to terminate such support.*

[12] The Original Article regulating this issue more detailed in Annex I was Art. 16, but after the amendment of the Additional Protocol I on November 30, 1993, it was renumbered as Art. 17. The text of the Article remained unchanged:

1. *The international special sign for works and installations containing dangerous forces, as provided for in Article 56, paragraph 7, of the Protocol, shall be a group of three bright orange circles of equal size, placed on the same axis, the distance between each circle being one radius, in accordance with Figure 5 illustrated below.*
2. *The sign shall be as large as appropriate under the circumstances. When displayed over an extended surface it may be repeated as often as appropriate under the circumstances. It shall, whenever possible, be displayed on flat surfaces or on flags so as to be visible from as many directions and from as far away as possible.*
3. *On a flag, the distance between the outer limits of the sign and the adjacent sides of the flag shall be one radius of a circle. The flag shall be rectangular and shall have a white ground.*
4. *At night or when visibility is reduced, the sign may be lighted or illuminated. It may also be made of materials rendering it recognizable by technical means of detection.*

2.1 The Environmental Modification Techniques Convention (ENMOD)

The basic scope of protection related to the ENMOD Convention is given in Art. 1:

> ...each State Party in this Convention undertakes not to engage in military or any other hostile use of environmental modification techniques having widespread, long-lasting or severe effects as the means of destruction, damage or injury to any other State Party.

Art. 2 clarifies the term "environmental modification technique" from Art. 1:

> ...refers to any technique for changing – through the deliberate manipulation of natural processes – the dynamics, composition or structure of the Earth, including its biota, lithosphere, hydrosphere and atmosphere, or of outer space.

More detailed examples of the environmental modification techniques are

> ...earthquakes; tsunamis; an upset in the ecological balance of a region; changes in weather patterns (clouds, precipitation, cyclones of various types, and tornadic storms); changes in the climate patterns; changes in ocean currents; changes in the state of ozone layer; and changes in the state of the ionosphere.[13]

As this treaty was drafted and adopted partially as a response to the USA defoliation campaigns in the Vietnam War, it prohibits the deliberate manipulation of natural processes (i.e. it prohibits environmental modification techniques) as a weapon during conflict. Although it is a common belief that it prohibits the use of the environment as a weapon, as opposed to Art. 35 (3) of Additional Protocol I, which protects the environment as a potential military objective, one author rightly pointed out that the object of protection in Art. 1 of the ENMOD is not the environment itself, but the prevention of "destruction, damage or injury to any other State Party".[14]

When analyzing the formulation used in Art. 1 of the ENMOD Convention, i.e. "widespread, long-lasting or severe" effects, there are less ambiguities when compared to the similar wording of Additional Protocol I, due to the fact that there is an official interpretation of these expressions in the "Understandings",[15] which were enclosed in the text of the Convention during the negotiating process and submitted to the General Assembly of the United Nations.[16]

The definitions in the "Understandings" are as follows:

(a) "Widespread": encompassing an area on the scale of several hundred square kilometres
(b) "Long-lasting": lasting for a period of a month, or approximately a season
(c) "Severe": involving serious or significant disruption or harm to human life, natural, economic resources or other assets

It is obvious that the threshold of the three expressions mentioned earlier is more precise and less demanding than that in Additional Protocol I. This conclusion

---

[13] See: *supra* 5 (p. 168).
[14] See: *supra* 10 (pp. 16–17).
[15] See: *supra* 5 (p. 417).
[16] See: *supra* 5 (p. 168).

results not only from the comparisons of the thresholds in the two documents, but also from the non-cumulative standard of protection originating from the word or (widespread, long-lasting or severe) instead of the word and (widespread, long-term and severe) used in Additional Protocol I. Another important difference between the two treaties is that the environmental protection stated in Additional Protocol I is granted over the duration of the armed conflict, while in the ENMOD Convention protection is granted both in peacetime as well as in times of armed conflict. Despite this, the ENMOD Convention is subject to serious criticism, since it protects only parts, objects or assets of the environment within the territory of the state parties; it is not applicable to many smaller states; it does not include some traditional methods of causing environmental harm; and finally it has a certain vagueness and several loopholes in the text.[17]

2.2 The Statute of the International Criminal Court

The armed conflicts in the 1990, especially the First Gulf War and the armed conflict which hit the countries of former Yugoslavia also had a significant impact on the development of the understanding of individual criminal responsibility with respect to the crimes committed against the environment.

Based on the relevant Art. 8(2)(b)(iv) of the ICC Statute the following is included as a crime:

> Intentionally launching an attack in the knowledge that such attack will cause.... widespread, long – term and severe damage to the natural environment which would be clearly excessive in relation to concrete and direct overall military advantage anticipated.

It seems obvious that the model for the wording of Art. 8(2)(b)(iv) of the ICC Statute are Art. 35 (3) and Art. 55 of Additional Protocol I. Further reference to the Additional Protocol I can also be made based on Art. 21 (1) (b) and Art. 31 (3) (c) of the ICC Statute. Although the meaning of the expression "natural environment" in this context is debatable, the view that this notion has to be understood broadly, seems the most appropriate in the context of the ICC Statute.[18] Compared to the one in Additional Protocol I, the damage threshold in Art. 8(2)(b)(iv) of the ICC Statute seems to be at least very similar with respect to the three conditions (widespread, long-term and severe), which have to be met cumulatively as is the case with Additional Protocol I. Nevertheless, it has to be borne in mind that the three elements have not been defined in the ICC Statute. Regarding the potential comparison with the ENMOD, the Understanding relating to article I of ENMOD expressly excludes this possibility by pointing out that its definition "is intended exclusively for this Convention".[19]

---

[17] See: *supra* 10 (p. 18).

[18] See: Peterson (2009, p. 329). According to Peterson, the notion of natural environment "refers to all ecological conditions in which humans naturally live, including artificially created, modified, or otherwise reclaimed nature."

[19] See: *supra* 16.

On the other hand, the possible meaning of "an attack" and the very high criteria related to the use of the principle of proportionality and understandings of military objective and military necessity (clearly excessive in relation to concrete and direct overall military advantage anticipated) lead one author in a recent publication to express very serious doubts about any possibility for its practical application.[20]

2.3 The Practice of the International Court of Justice

In the ICJ's Advisory opinion in the Nuclear Weapon Case[21] there are several important paragraphs with respect to environmental protection in times of armed conflict. Although the ICJ did not find that there is a treaty obligation which:

> [...] could have intended to deprive a State of the exercise of its right of self-defence under international law because of its obligations to protect the environment. Nonetheless, States must take environmental considerations into account when assessing what is necessary and proportionate in the pursuit of the legitimate military objectives. Respect of the environment is one of the elements that go to assessing whether an action is in conformity with the principles of necessity and proportionality.[22]

In addition, the ICJ further recognizes that Art. 35 (3) and Art. 55 of Additional Protocol I taken together *"embody a general obligation to protect the natural environment against widespread, long-term and severe environmental damage [...]"*.[23] In the next paragraph the ICJ reiterates the text of the General Assembly resolution 47/37 dated 25 November 1992, which establishes that *"the destruction of the environment, not justified by military necessity and carried out wantonly, is clearly contrary to existing international law"*.[24] In his dissenting opinion concerning the same case, judge Weeramantry reminds, among other things, that in the draft articles on "State responsibility" of the International Law Commission the massive pollution of the atmosphere or of the oceans has been classified as an international crime.[25]

In 1999, the ICJ also received a complaint of Serbia and Montenegro (at the time Federal Republic of Yugoslavia) against ten states of the North Atlantic Treaty

---

[20] See: *supra* 18 (p. 10). After thoroughly explaining all shortcomings of Art. 8(2)(b)(iv) of the ICC Statute, I. Peterson concludes that: "...it can be assumed that Article 8(2)(b)(iv), second alternative, of the ICC Statute merely pays lips to environmental concerns, without creating the risk that anyone will be prosecuted for its particular offence."

[21] The ICJ Advisory opinion on the Legality of the Threat or Use of Nuclear Weapons, July 8, 1996, available at: http://www.icj-cij.org/docket/files/95/7495.pdf. Accessed on 23 Nov 2009.

[22] See: *Id*. Para 30. (p. 242).

[23] See: *Id*. Para 31. (p. 242).

[24] See: *Id*. Para 32. (p. 242).

[25] This separate opinion is available at: http://www.icj-cij.org/docket/files/95/7521.pdf. Accessed on 23 Nov 2009, p. 502.

Organization (NATO) alleging a breach of the obligation not to resort to the use of force. Among other things, the allegations were based on the grounds of breaching the obligation not to cause considerable damage, whereas the case against the United States of America related to the alleged breach of its obligation not to use prohibited weapons and not to cause far-reaching health and environmental damage. Ultimately, the cases were dismissed on the basis of lack of jurisdiction.[26]

2.4 Other Relevant Legal Rules on Armed Conflict

Before mentioning other relevant legal rules on armed conflict relating to environmental protection, one general principle of humanitarian law deserves special attention: the Martens clause.[27] It was originally included in the Preamble of the IV Hague Convention and today it is part of the 1949 Geneva Conventions, the 1977 Additional Protocols I and II and the 1980 Conventional Weapons Convention.[28] Today the Martens clause is undoubtedly an element of customary international law and its principles are applicable to environmental protection in armed conflict.[29]

As to other relevant legal rules of armed conflict also relating to environmental protection, if observed chronologically, the first would be the prohibition of the use of poison or poisoned weapons established at the Hague Conventions of 1899 and the 1907 Hague Convention on Laws and Customs of War on Land.[30] The 1925 Geneva Gas Protocol[31] is relevant for the prohibition of chemical warfare, as well as the use of bacteriological methods of warfare. Furthermore, high importance should also be

---

[26] These are the following cases: Legality of Use of Force (Serbia and Montenegro against Belgium), Legality of Use of Force (Serbia and Montenegro against Canada), Legality of Use of Force (Serbia and Montenegro against France), Legality of Use of Force (Serbia and Montenegro against Germany), Legality of Use of Force (Serbia and Montenegro against Italy), Legality of Use of Force (Serbia and Montenegro against Netherlands), Legality of Use of Force (Serbia and Montenegro against Portugal), Legality of Use of Force (Serbia and Montenegro against Portugal), Legality of Use of Force (Yugoslavia against Spain), Legality of Use of Force (Serbia and Montenegro against United Kingdom), Legality of Use of Force (Yugoslavia against United States of America). The cases are available on: http://www.icj-cij.org/docket/index.php?p1=3&p2=2. Accessed on 23 Nov 2009.

[27] Martens clause: "Until a more complete code of laws of war has been issued, the High Contracting Parties deem it expedient to declare that, in cases not included in the Regulations adopted by them, the inhabitants and the belligerents remain under the protection and the rule of the principles of the law of nations, established among civilised peoples, from the laws of humanity, and the dictates of the public conscience".

[28] Convention on Prohibitions or Restrictions on Use of Certain Conventional Weapon which may be deemed to be Excessively Injurious or to have Indiscriminate Effects (with Protocols I and II (1342) UNTS 137.

[29] More on Martens clause in relation to protection of environment in armed conflict see in: supra 4 (p. 13); supra 10 (pp. 21–22).

[30] Schindler and Toman (1988, pp. 63–98).

[31] See: supra 30. Protocol for the Prohibition of the Use in War of Asphyxiating, Poisonous or Other Gases, and of Bacteriological Methods of Warfare (pp. 115–127).

placed on the 1972 Bacteriological Convention,[32] which was amended by a Final Declaration in 1986. This instrument makes possession and use of bacteriological weapons illegal. Other milestones in the development of this field include the acceptance of the 1993 Convention on Prohibition of the Development, Production, Stockpiling and Use of Chemical Weapons and Their Destruction as well as the establishment of the Organisation for the Prohibition of Chemical Weapons in 1997.[33]

On the other hand, the 1980 Conventional Weapons Convention contains general principles regarding the use of conventional weapons and includes the following protocols: Protocol I on non-detectable fragments; Protocol II on mines, booby traps and other devices; Protocol III on incendiary weapons; Protocol IV on blinding lasers. The 1996 Amended Protocol extended protection, but further development resulted in establishing a new treaty: the 1997 Convention on the Prohibition of Use, Stockpiling, Production and Transfer of Anti-Personnel Mines and on their Destruction (Ottawa Treaty),[34] which introduced total prohibition of anti-personnel mines.

2.5 Other Relevant Peacetime Legal Rules

When trying to analyze other environmental treaties and their potential relation to the environmental protection, one has to bear in mind that this problem is closely related to certain general issues such as the applicability of such treaties concluded in peacetime for times of armed conflict.[35] After an extensive analysis of peacetime treaties relevant for environmental protection in times of armed conflict, Vöneky[36] distinguishes five categories of treaties binding state parties in times of armed conflict[37]:

(a) Treaties that expressly provide for continuance during war
(b) Treaties that are compatible with the maintenance of war
(c) Treaties creating an international regime or status
(d) Human rights treaties
(e) *Ius cogens* rules and obligations *erga omnes*

---

[32] Convention on the Prohibition of Development on the Prohibition of Development, Production and Stockpiling of Bacteriological (Biological) and Toxin Weapons and on their Destruction (1972) 1015 UNTS.

[33] See more: http://www.opcw.org. Accessed on 23 Nov 2009.

[34] Convention on the Prohibition of Use on the Prohibition of Use, Stockpiling, Production and Transfer of Anti-Personnel Mines and on their Destruction (1997), 2056 UNTS 211.

[35] See: Sands (2003).

[36] Silja (2000, pp. 190–225).

[37] See: *Id.* (p. 198).

(a) This category of treaties is not very numerous, as such treaties cannot be suspended or terminated during the wartime, except when expressly provided.[38]

One of the treaties included in this category is the 1954 Convention for the Prevention of Pollution of the Sea by Oil (OILPOL).[39]

(b) The suspension or termination of this kind of peacetime treaties, i.e. environmental multilateral treaties whose execution is compatible with the maintenance of war, is not allowed and they continue to apply during war even between belligerent states.[40]

(c) The characteristic of these treaties is that they create a kind of territorial order which serves the interest of the international community, *"such as are treaties providing for the demilitarization or neutralization of zones or the internationalization of waterways"*.[41]

(d) Although there is a common view that the human rights treaties in general continue to apply during war, Vöneky considers that *"it seems that only massive and severe damage to the environment constitutes a violation of human rights"*.[42] The protection of the environment in human rights treaties is naturally a clear expression of the *"anthropocentric value approach"* (see Sect. 1.1).

(e) Albeit it looks like it is almost impossible to find any norm of international environmental law of ius cogens character during armed conflict, Vöneky concludes, after a thorough analysis of the International Law Commissions Draft Principles on State Responsibility, that *"at least the prohibition of massive pollution of the environment is a peremptory norm of international law and an obligation with effect erga omnes"*.[43]

3 The Proposals for Enhancing the Effectiveness of Environmental Protection in Times of Armed Conflict

Numerous proposals for the enhancement of the effectiveness of environmental protection in times of armed conflict may be grouped in several different categories:

(a) The first group includes proposals which are familiar with those for the enhancement of the effectiveness of the law of armed conflict in general. As a first interim measure, Schmitt proposes that "significant resources should be dedicated

---

[38] See: *Id.*
[39] See: *Id.*
[40] See: *Id.* (p. 199).
[41] See: *Id.*
[42] See: *Id.* (pp. 200–201).
[43] See: *Id.* (pp. 202–203).

to exploring the environmental consequences of armed conflict".[44] However, his following proposals are connected with the broader inclusion of the environment as part of the training of military lawyers and with the encouragement of the non-parties, especially the most powerful states, to the ratification of the relevant treaties.[45]

(b) The second group of proposals refers to a different understanding of the contcept of environment. Some of the most important proposals with respect to this position stem from the understanding that "the environment is indivisible: that also the environment constitutes an indivisible entity composed of delicately interdependent ecological systems, parts, objects, and assets."[46] "Interdependent (including transboundary) features of environmental processes"[47] are also related to this, especially the features and consequences of armed conflict. Moreover, the understanding of these processes lead to considering the development of international legal concepts which could be suitable for these purposes as well. One of the most interesting proposals suggests extending "the concept of the common heritage of mankind" to the environment.[48] As this last concept seems somewhat too radical if referred to the environment as a whole, some authors consider the "concept of common concern" with respect to the environment as a possible solution.[49] However, the exact content of the latter concept is not completely clear.

(c) The third group relates to proposals for introducing new treaties and rules for environmental protection in times of armed conflict. Some of these proposals included the initiative for drafting the "Fifth Geneva Convention for the Protection of Environment in Time of Armed Conflict".[50] On the other hand, one author does not consider this idea to be the best possible solution and instead, he proposes his own text entitled the "Convention on the Protection of the Environment During Armed Conflict".[51] The text of the later Convention protects the environment in times of non-international armed conflict as well. The endeavours of certain international organizations in this field should also be mentioned, first of all the "Guidelines for military manuals and instructions on the protection of the environment in times of armed conflict (Guidelines)" by the International Committee of the Red Cross (ICRC).[52] The Guidelines are established as a follow-up to the International Conference for the Protection of

---

[44] See: *Supra* 4. (p. 23).
[45] See: *Supra* 4. (p. 23).
[46] See: *Supra* 10. (pp. 32–33).
[47] See: *Id*. (p. 34).
[48] See: *Id*. (p. 37); see also: *supra* 36 (p. 206).
[49] See: *Supra* 10. (p. 37).
[50] See: *Id*. (p. 8).
[51] Talbot (2005, pp. 145–185).
[52] For the text of the guidelines see: http://www.icrc.org/Web/Eng/siteeng0.nsf/html/57JN38. Accessed on 25 Nov 2009. More about relation of Guidelines with the ENMOD Convention see: Roman (2007, p. 327).

War Victims in 1993 and in the view of the ICRC they represent "a summary of the existing applicable international rules which must be known and respected by members of armed forces".[53]

4 The Protection of Water Resources in Times of Armed Conflict

On the one hand, the international legal rules related to environmental protection in general developed significantly in the last few decades, on the other hand similar rules started to develop as well for certain particular fields, such as in the case of the protection of water resources in times of armed conflict. If we try to offer a short overview of these specific rules related to water resources, we should mention first that the basic rules regulating this field are various rules of treaty and customary international law analyzed in the previous chapters of this paper, which are applicable to water resources as part of the (natural) environment.

The development of special rules in relation to water resources in times of armed conflict started almost in parallel with the development of modern international law in general. It is interesting that one of the first scholars of international law, Emmerich de Vattel, in his famous "Treatise on the Law of Nations" in 1758 mentioned *"a still more general unanimity prevails in condemning the practice of poisoning waters, wells and springs, because (say some authors) we may thereby destroy innocent persons."*[54] However, the most systematic work on the consolidation of the rules related to the protection of water resources comes much later, with the work of the International Law Association (ILA) and in particular with the ILA's Committee on the International Water Resources Law. The result of the work of this Committee was the adoption of the Resolution on the Protection of Water Resources and Water Installations in Times of Armed Conflict in 1976 (1976 Resolution).[55] The Preamble of the 1976 Resolution mentions, among other things, *"the lack of specific rules in international law for the protection of water and water installations against damage or destruction in times of armed conflict"*,[56] but also that the 1976 Resolution presents guidelines for the elaboration of such rules. The last statement is of particular importance, because it shows the degree of awareness with respect to the soft law value of this document, although some parts of it are undoubtedly included in customary international law. In any case, a short analysis of the Resolution consisting of eight articles seems appropriate.

---

[53] See: http://www.icrc.org/Web/Eng/siteeng0.nsf/html/57JN38. Accessed on 25 Nov 2009.

[54] ILA, Report of the fifty-seventh conference, Madrid 1976 (resolution of adoption, pp. xxxiv–xxxvi; Report of the Committee on the International Water Resources Law, Part II – The Protection of Water Resources and Water Installations in Times of Armed Conflict, pp. 234–237; Rapporteur F.J. Berber). In Bogdanovi (2001, p. 225).

[55] See: *Id.* (pp. 233–242).

[56] See: *Id.* (p. 233).

Article I

> Water which is indispensable for the health and survival of the civilian population should not be poisoned or rendered otherwise unfit for human consumption.

The first Article represents an elaboration of the opinion mentioned by de Vattel and even before, by Alberico Gentili in his *"De Jure Belli"* from 1588/89, with respect to this rule as an established part of customary international law.[57] The rule is a clear demonstration of the anthropocentric approach in environmental protection and at first glance it does not seem questionable. However, on the other hand, the interpretation of this Article raises a number of uncertainties. The most important of them are related to the use of atomic weapons the use of which would certainly mean at least a serious contamination of water resources.[58] In our view, there is no doubt that this article is applicable in case of use of atomic weapons, although for water resources the use of these weapons presents only one negative effect regarding its destructive strength on the environment in general. When concerning the other doubt, i.e. the applicability of the Article on the combatants as well as on the civilian population, it seems natural that the Article also protects the combatants. Another question, to which there is still no definite answer, is *"whether the extension of the rule to all means of making water unusable for human consumption can be applicable to measures directed against combatants only"*.[59] In other words, it is not quite clear whether making water unfit for human consumption, when directed (if possible at all) only against combatants, is still allowed as a legitimate military measure.[60]

Article II

> Water supply installations which are indispensable for the minimum conditions of survival of the civilian population should not be cut off or destroyed.

This Article does not look to raise serious doubts, although there is a practical problem with the distinction of water installations serving (only) for the civilian population from those serving only for the military.

Article III

> The diversion of waters for military purposes should be prohibited when it would cause disproportionate suffering to the civilian population or substantial damage to the ecological balance of the area concerned. A diversion that is carried out in order to damage or destroy the minimum conditions of survival of the civilian population or the basic ecological balance of the area concerned or in order to terrorize the population should be prohibited in any case.

The diversion of a river for military purposes is a very old military measure which, if carried out in a way not to cause disproportionate suffering to the civilian population or substantial damage to the environment, can be a legitimate military

---

[57] See: *Id.* (pp. 233–234).
[58] See: *Id.* (p. 234).
[59] See: *Id.*
[60] See: *Id.*

measure.[61] However, the question is at which moment the negative consequences described in the second sentence of this Article arise for the civilian population or for the environment itself. Obviously, this Article merges the anthropocentric with the intrinsic value approach of environmental protection.

Article IV

> The destruction of water installations, such as dams and dykes, which contain dangerous forces, should be prohibited when such destruction might involve grave dangers to the civilian population or substantial damage to the basic ecological balance.

The prohibition of the destruction of water installations, with the examples of dams and dykes is almost simultaneously regulated in Art. 56 of Additional Protocol I (see Sect. 1.1.) in a more detailed way. It is important that in accordance with the interpretations of this Article, there are possible exceptions from the basic rule of this Article, i.e. situations when it would be possible to take military action against water installations.[62] The first situation would be the case of unintentional destruction or damage as a result of the legal bombardment of neighbouring military installations.[63] On the other hand, the development of military equipment in the last few decades with very precise targeting possibilities should limit such occurrences to the lowest possible level. The second possibility is related to potential attacks on modern, multi-purpose hydro-electrical installations, which in addition to its military significance have an important function in the production of electric power for civilian purposes.[64] It is not clear whether the protection granted by virtue of this Article expands completely to these multifunctional hydro-electrical installations as well.

Article V

> The causing of flood as well as any other interference with the hydrological balance by means not mentioned in Arts. II to IV should be prohibited when it involves grave dangers to the civilian population or substantial damage to the ecological balance of the area concerned.

The causing of floods also does not represent a new means of waging war.[65] The prohibition in this Article can be interpreted not only as a prohibition of destruction which can be performed with the purpose of causing flood. The expression "as any other interference with the hydrological balance not mentioned in Arts. II to IV" seems broad enough to encompass environmental modification techniques in the sense accepted in the ENMOD Convention (see Sect. 1.2.). The question remains however, whether it can be claimed or not that this prohibition is already part of customary international law.[66]

---

[61] See: *Id.* (p. 235).

[62] See: *Id.* (p. 236).

[63] See: *Id.* This article also establishes the balance between the anthropocentric and intrinsic value approach in environmental protection.

[64] See: *Id.*

[65] See: *Id.* (p. 237).

[66] See: *Id.* (p. 238).

Art. VI

1. The prohibitions contained in Arts. I to V should be applied also in occupied enemy territories.
2. The occupying power should administer enemy property according to the indispensable requirements of the hydrological balance.
3. In occupied territories, seizure, destruction or intentional damage to water installations should be prohibited when their integral maintenance and effectiveness would be vital to the health and survival of the population.

It is obvious that this Article is formulated under the influence of the Hague Conventions, first of all Art. 55 and 56 of the Fourth Hague Convention from 1907,[67] although the Hague Conventions themselves do not mention water installations explicitly.

Article VII

The effect of the outbreak of war on the validity of treaties or of parts thereof concerning the use of water resources should not be termination but only suspension. Such suspension should take place only when the purposes of the war or military necessity imperatively demand the suspension and when the minimum requirements of subsistence for the civil population are safeguarded.

The validity of treaties in general in time of armed conflict has already been the subject of Sect. 3.1 of this paper. The intention of this Article is plainly the limitation of the potential termination of treaties or parts thereof concerning the use of water resources by providing the possibility of suspension of such treaties in very determinable circumstances of the second sentence of this Article. It seems that this possibility still remains lex ferenda, rather than lex lata, of course if it is not explicitly provided in a specific treaty.[68]

Article VIII

1. It s hould be prohibited to deprive, by the provisions of a peace treaty or similar instrument, a people of its water resources to such an extent that a threat to the health or to the economic or physical conditions of survival is created.
2. When as the result of the fixing of a new frontier, the hydraulic system in the territory of one State is dependent on works established in the territory of another, arrangements should be made for the vital needs of the people.

The provisions in this Article are related to the situation when in post-conflict times peace treaties or similar instruments are created. It is worth mentioning that in such situations, as the Rapporteur rightly reminded,[69] Art. 53 of the Vienna Convention on the Law of Treaties[70] which states that a treaty is void if, at the time of its conclusion,

---

[67] Art. 55. Convention (IV) respecting the Laws and Customs of War on Land. Signed at The Hague, 18 October 1907, *supra* 32 (pp. 91–92).

[68] We agree with the conclusion of the Rapporteur on this problem, i.e.: "A very detailed study of a great number of water treaties would have to be made in order to find out and formulate the various implications of the above mentioned very general rule on the multiple various classes of water treaties." *Id.* (p. 241).

[69] See: *Id.* (p. 242).

[70] Vienna Convention on the Law of Treaties (1969), 1155 UNTS 331.

it conflicts with a peremptory norm of general international law, applies. In the context of this Article, this means that a treaty can be regarded as void, if its content is directed towards the deprivation of people of their water resources in situations described in this Article. In addition to this, certain provisions in peace treaties concluded after the First World War were subject to the *Rapporteur*'s scrutiny in this context.[71]

At the end of a long process (which lasted from 1990 until 2004), the Water Resources Committee (WRC) adopted in 1999 the consolidated rules relating to international water resources – the "Campione Consolidation of the ILA Rules on International Water Resources". In relation to that, Chapter VII (Art. 37 – 44) includes all eight articles of the 1976 Resolution, almost unchanged when compared to the 1976 Resolution,[72] which ultimately served for the development of the ILA Berlin Rules in 2004, at the 71st ILA Conference.

4.1 Other Relevant Rules Related to the Protection of Water Resources in Times of Armed Conflict

The International Law Commission (ILC) also contributed to the development of legal rules related to the protection of water resources in times of armed conflict. In 1994, the ILC published its draft articles on the Law of Non-navigational Uses of International Watercourses (1994 Draft articles).[73] Article 29 is related to the situation of armed conflict.

Art. 29 International watercourses and installations in times of armed conflict:

> International watercourses and related installations, facilities and other works shall enjoy the protection accorded by the principles and rules of international law applicable in international and internal armed conflict and shall not be used in violation of those principles and rules.

Based on the official Commentary of this rule,[74] it is evident that this provision does not introduce any new rule in this field, but reiterates existing ones. However, it should be emphasized that this Article seems to be applicable in both international and internal armed conflict, which is not easily established for many legal sources related to environmental protection in times of armed conflict. In 1997, the General Assembly of the United Nations adopted the Convention on the Law of Non-navigational Uses of International Watercourses (1997 Convention).[75] This Convention is modelled on the 1994 Draft articles of the ILC and Art. 29 is literally the same as the one included in the 1994 Draft articles. The 1997 Convention is not yet in force.

---

[71] See: *supra* 56 (p. 242).

[72] For Campione Consolidation see: Slavko Bogdanovi (1999, pp. 141–201).

[73] International Law Commission (1994), Draft articles on the law of non-navigational use of international watercourses and commentaries thereto and resolution on transboundary confined groundwater, *Yearbook of the International Law Commission*, vol. II, Part Two.

[74] See: *Id.* (p. 131).

[75] Convention on the Law of the Non-navigational Uses of International Watercourses, Official Records of the General Assembly, Fifty-first Session, Supplement No. 49 (A/51/49).

5 Conclusion

Although the international rules on environmental protection developed extensively in the last few decades, the most important treaties and related practice still include Art. 35 (3) and Art. 55 of Protocol I Additional to the 1949 Geneva Conventions, the ENMOD Convention and the relevant practice of the ICJ. By including Art. 8 (2)(b)(iv) of the Statute of the ICC new hope is given with respect to the development of the individualization of the crimes against the natural environment. However, at the same time this Statute is subject to serious criticism due to the restrictiveness of its threshold. The possibilities relating to the application of peacetime environmental treaties for the times of armed conflict have not yet been exhausted and further development in this direction is expected and desirable, for which different proposals have been mentioned in Sect. 4.1. The analysis of the relevant rules related to the protection of water resources and water installations in times of armed conflict, first of all the 1976 Resolution, shows tendencies of this specific field of protection to elaborate and slowly develop its own rules which are principally in line with the general rules. The further development of these special rules should be welcomed as they develop the protection system in a more detailed manner taking into account the peculiarities of a certain field, in this case related to the protection of water resources and water installations in times of armed conflict.

References

Bogdanović S (1999) International law association rules on international water resources – Pravila Udruženja za međunarodno pravo o međunarodnim vodenim resursima. Yugoslav Association for Water Law, Yugoslav Branch of ILA, Prometej and European Center for Peace and Development in Belgrade, Novi Sad

Bogdanović S (ed) (2001) International law of water resources contribution of the international law association (1954–2000). Kluwer Law International, the Hague

Peterson I (2009) The natural environment in times of armed conflict: a concern for international war crimes law. Leiden J Int Law 22(2):325–343

Roman R (2007) Protection of the environment during armed conflict. Missouri Environ Law Policy Rev 14:323–338

Sands P (2003) Principles of international environmental law. Cambridge University Press, Cambridge

Sandoz Y, Swinarski C, Zimmermann B (eds) (1987) Commentary on the additional protocols of 8 June 1977 to the Geneva conventions of 12 August 1949. Martinus Nijhof, Geneva

Schindler D, Toman J (1988) The laws of armed conflicts a collection of conventions, resolutions and other documents, 3rd edn. Martinus Nijhoff, Dordrecht

Schmitt M (2000) Humanitarian law and the environment. Denver J Int Law Policy 28(3):265–323

Silja V (2000) Peacetime environmental law as a basis of state responsibility for environmental damage caused by war. In: Jay E Austin, Carl E Bruch (eds) The environmental consequences of war, Cambridge University Press, Cambridge, pp 190–225

Talbot JE (2005) The international law of environmental warfare. active and passive damage during armed conflict. Vanderbilt J Trans Law 38(1):145–185

Verwey D (1995) Protection of the environment in times of armed conflict: in search of a new legal perspective. Leiden J Int Law 8(1):7–40

Documents

Convention on Prohibitions or Restrictions on Use of Certain Conventional Weapon which may be deemed to be Excessively Injurious or to have Indiscriminate Effects (with Protocols I, II and III), 1342 UNTS 137

Convention on the Prohibition of Development, Production and Stockpiling of Bacteriological (Biological) and Toxin Weapons and on their Destruction (1972) 1015 UNTS

Convention on the Prohibition of Use, Stockpiling, Production and Transfer of Anti-Personnel Mines and on their Destruction (1997), 2056 UNTS 211

Convention on the Law of the Non-navigational Uses of International Watercourses, Official Records of the General Assembly, Fifty-first Session, Supplement No. 49 (A/51/49)

Convention on the Prohibition of the Use of Environmental Modification Techniques (1977) 1108 UNTS 151

International Law Commission (1994) Draft articles on the law of non-navigational use of international watercourses and commentaries thereto and resolution on transboundary confined groundwater, Yearbook of the International Law Commission, vol. II, Part Two

Protocol Additional to the Geneva Conventions of 12 August 1949, and Relating to the Protection of Victims of International Armed Conflicts (1977) 1125 UNTS

Rome Statute of the International Criminal Court (1998) 2187 UNTS 3

Vienna Convention on the Law of Treaties (1969) 1155 UNTS 331

Internet Sources

http://www.icj-cij.org/docket/files/95/7495.pdf. Accessed on 23 Nov 2009
http://www.icj-cij.org/docket/files/95/7521.pdf. Accessed on 23 Nov 2009, p 502
http://www.icj-cij.org/docket/index.php?p1=3&p2=2. Accessed on 23 Nov 2009
http://www.icrc.org/Web/Eng/siteeng0.nsf/html/57JN38. Accessed on 25 Nov 2009
http://www.opcw.org. Accessed on 23 Nov 2009

Food Security and Eco-terrorism Impacts on Environmental Security Through Vulnerabilities

Hami Alpas and Taylan Kiymaz

Abstract Food security requires appropriate agricultural management and utilization of natural resources and eco-systems, as well as good governance and sustainable political systems. Food security is directly affected by climate change effects that lead to concerns in rural livelihoods. Bio-energy developments present both opportunities and challenges for socioeconomic development and the environment. In that sense, bio-energy solutions should strive to be environmentally sensitive and have a positive social impact. On the other side, trade policy enforced via World Trade Organisation (WTO) is expected to play a role in mitigating and adapting to global climate change by increasing incentives to use the most energy efficient environmental goods and services.

The recent food and financial crises developed from different underlying causes but intertwined in complex ways through their implications for financial and economic stability, food security, and political security. As the majority of the poor in the world are considered to be depended on agriculture, the severeness of the climate change effects may lead to food system risks and more of the societal and political risks can be incurred in the future. The potential of food price volatility and climate change leading a rise in food insecurity among the poor groups is significant, and thus, can be expected to bring social disturbances and terrorism in the short to long term. Food security is a hot topic. Therefore, its disruption via environmental breakdown is an obvious cause for terrorism. Although the biotechnology revolution is very relevant to the problems of food security, poverty reduction, and environmental conservation in the developing world, it raises many questions relating to ethics, intellectual property rights, and bio-safety. Some policy alternatives for environment friendly support of food security consist in increasing productivity on the non-forest fertile soils and in animal production systems or to reduce postharvest losses, providing greater incentives for agriculture to use water more efficiently, promoting larger

H. Alpas (✉)
Department of Food Engineering, Middle East Technical University, Ankara 06531, Turkey
e-mail: imah@metu.edu.tr

T. Kiymaz
State Planning Organization, Ankara, Turkey

investments in agricultural research to raise production with environment friendly techniques. Global environmental change (GEC) will have serious consequences for food security, particularly for more vulnerable groups. Adapting to the additional threats to food security arising from major environmental changes requires an integrated food system approach, not just a focus on agricultural practices. In this respect, vulnerability assessment could help to address food supply-chain security by determining the selection of countermeasures and emergency responses.

Keywords Food security • Ecoterrorism • Environmental security • Climate change

1 Introduction

Ensuring food security today not only requires appropriate agricultural management and utilization of natural resources and eco-systems, but also good governance and sustainable political systems. Land and water constraints, underinvestment in rural infrastructure and agricultural innovation, lack of access to agricultural inputs and weather disruptions are impairing productivity growth and the needed production response. These factors, combined with sharp increases in food prices in recent years, have added concerns about the food and nutrition situation of people around the world, especially the poor in developing countries. Several factors have contributed to these unprecedented food price increases: climate change, rising energy prices and subsidized bio-fuel production, income and population growth (Von Braun et al. 2008a), and also speculations in stock markets.

Food security is directly affected by climate change effects that lead to concerns in rural livelihoods. Food safety problems indirectly affect food security but are directly affected by climate change effects. It has long been recognized that social conflicts increase food insecurity, but it should also be pointed out that food insecurity can be a key source of conflict. Energy-security objectives that have led to subsidized expansion of bio-fuel production also contribute to food insecurity. Thus, food security, energy security, environmental security and political security, and their associated risks are linked and bring about important trade-offs. Slow increases in world food production, rapid decrease in stocks and declining rates of yield growth in main food crops endanger world food security.

The poorest usually suffer silently for a while, but the middle class typically has the ability to organize, protest, and lobby early on. Since early 2007, increasing food costs and general living costs have led to social and political unrests in about 60 countries, with some experiencing multiple occurrences and a high degree of violence. While this unrest has occurred in countries with low performance in governance, other countries have also been affected. In this study, some of the mentioned factors threatening food security are linked with environmental security issues considering eco-terrorism as a side effect. In that manner, the recent global situation is analyzed and a policy framework is foreseen for the future, putting more emphasis on the environment.

2 Bio-energy Policies

Without considering global food security concerns, most resource-rich countries are encouraging bio-energy development via support policies. They have at least one of the following policy objectives: to increase energy security, to diversify energy sources, to support rural development, to reduce the impact of energy use on climate change or more generally to improve the environment.

Bio-energy developments present both opportunities and challenges for socio-economic development and the environment. These developments have a number of potential impacts on forests and the rural poor population. In developing countries, the impact of bio-energy on poverty alleviation will depend on the opportunities that are presented for agricultural development.

The use of staple crops for energy production may lead to food insecurity. Energy production can be expected to bring land conflicts if access to land will be restricted. The environmental impacts of these developments are uncertain and may vary in each case. The development of bio-energy is likely to have significant impact on the forest sector directly through the use of wood for energy production.

In most countries, current bio-energy support policies have sometimes conflicting objectives that require careful policy choices. These policies are considered to create good results for energy security and rural development, whereas they are likely leading to food price implications and natural resource impacts. Increased consumption of bio-energy may result in increased competition for land that has potential to impact agriculture and forestry and could negatively affect the poor, such as through changes in access to resources (Anonymous 2009a).

According to the OECD, bio-energy contributes an increasing share to total primary energy supplies of both OECD and non-OECD countries, although overall shares tend to remain relatively small in most countries, particularly when it comes to energy based on agricultural biomass. The implications of growing production and use of bio-energy for agricultural production systems are, however, of great interest, both considering historical developments and future changes. In the context of liquid bio-fuels, this growing industry and the policy support measures behind this growth can result in increased use of agricultural produce for non-food purposes, higher agricultural commodity prices at regional and international levels, and – in a regionally differentiated manner – expanded land use for agricultural production with potentially negative effects for the environment (OECD 2008).

With respect to solid biomass, wood pellet use is expected to increase in developed and some developing countries. This growth in demand will not be met without imports especially in the European Union (EU), including imports from the tropics. This could increase pressures on land and for local populations if it is not under sustainable production schemes.

Bio-energy solutions should strive to be environmentally sensitive and have a positive social impact. There appear to be more opportunities for this with regard to solid biomass than liquid bio-fuels (based on current feedstocks and production methods), which tend to have larger environmental risks and mixed benefits for the poor.

Although studies are limited and would benefit from further work to clarify impacts, it is estimated that use of biomass resulted in average green house gas (GHG) savings ranging from 50% to 89% when compared to use of fuel oil for heating, and ranging from 31% to 84% compared with use of gas for heating (OECD 2009).

In producer countries, it is important to balance production targets with environmental and social concerns, including food security. The tradeoffs of bio-energy production should be carefully considered in order to determine the correct feedstock for a particular location to balance rural development priorities. It makes a lot of sense to establish some regional criteria within countries that have national biofuel support policies to differentiate areas with very low environmental risks and areas of high risks. The support policies for feed stocks should also guide investments into raw materials that meet best practices for environmental, social and climate change considerations.

As a result of various initiatives that are being developed to reduce carbon emissions and environmental degradation (including payments for environmental services, carbon markets and bio-energy developments), new demands are being placed on environmental goods and services, and lands (including forests) worth monetary value. These activities become more attractive to investors. However, this can result in reduced access to the land or reduced food security (Anonymous 2009b).

3 World Trade Organization (WTO) and Doha Round Negotiations

The function of the WTO is to offer a mechanism to countries to negotiate collective disciplines on policies that impose negative spillovers. Two important examples of the "public goods" that the system provides are a greater stability of prices in world food markets and an enhanced environmental sustainability.

Trade policies can have adverse consequences on the environment. Agricultural support programs have led to the use of production methods that are excessively polluting; as have fossil fuel subsidies. Subsidies have also encouraged the overexploitation of natural resources. Doha can help limit the negative environmental externalities. Trade policy has a role to play in mitigating and adapting to global climate change by increasing incentives to use the most energy efficient environmental goods and services (Hoekman et al. 2009).

Since the 1980s, subsidies become a large component of farmers' incomes and their land use decisions. The way in which these subsidies are allocated plays a major role in shaping land use patterns, particularly in the EU and the US, and therefore it has important impacts on the environment in rural areas. These countries utilize agri-environmental programs to encourage the provision of environmental amenities and to reduce negative environmental effects associated with agriculture. On the other hand, even economically efficient means of support like green box subsidies (permitted measures of support given as defined in the WTO jargon) may provide support for activities that are damaging the environment, although these types of subsidies are closely targeted at the achievement of concrete environmental goals. In the EU, the 2003 decoupling reform was effective in removing the incentive to overproduce, while also establishing several schemes with explicit environmental objectives. However, such environmental programmes are only effective if they have clear goals expressed in terms of measurable outcomes and target. In the US, it is argued that green box payments have perpetuated environmental problems in that they encourage production on marginal lands, for example through regular disaster assistance or some farm credit and stimulate maintaining production rather than retiring land in environmentally fragile areas (Ortiz et al. 2009).

Increasing climate stresses and the retreat of the private sector insurance industry from covering losses caused by catastrophic natural events will lead to increasing calls for national and local governments to step in. Included in the green box measures, most governments already operate public sector insurance programs for major risks if there is no private sector coverage, such as for crop loss, flood (FAO 2008). These kinds of programs would help resource poor rural population to be compensated for their unexpected losses and continue their livelihood in the homeland.

With the prolonging new multilateral disciplines that Doha Round is expected to bring, countries have been less able to resist protectionist pressures after food crisis in 2007–2008 period. The recent reintroduction of export subsidies for dairy products by the EU and the US is a good example. Climate change legislation has also been tabled in the US and EU that could lead to trade restrictions against countries that do not take measures that is perceived to be adequate. WTO rules need to be clearer on the scope for environmentally-motivated trade action (Hoekman et al. 2009).

4 Food Prices and Environmental Approach

International food grain and fertilizer prices have declined significantly from the highs experienced during the first 6 months of 2008. International prices of most grains and fertilizer have fallen significantly as improved grain supplies resulting from favourable harvests have boosted global stocks, and as fertilizer production increased. This, along with weaker demand for internationally traded food commodities under a global recession, and lower crude petroleum prices (that lowered the demand for ethanol based bio fuels, and for oil based fertilizers) has allowed prices to decline. Overall, concerns about the adequacy of global food supplies have subsided, and many of the export bans and high export taxes that were put in place during the food price spike of 2008 have either been eliminated or substantially reduced.

Although international prices have declined from their peaks last year, world grain prices in November 2009 were about 40% higher than 2006 average. Prices for commonly user fertilizers, such as urea and DAP, have declined significantly from their peak, with September 2009 prices being about 20% respectively higher than the prices at the start of 2006.

Despite declines in global food prices from the highs of mid-2008, depreciating exchange rates from the financial crisis have kept local prices of food imports high in many developing countries. In East Africa, grain prices in some areas are as high due to drought as they were during the peak of the food crisis. Lower remittances have further reduced household purchasing power. Less private investment, higher local costs of borrowing, and reduced government revenue and spending, all reduce the capacity of households to respond (Anonymous 2009a).

Although the food and financial crises developed from different underlying causes, they are becoming intertwined in complex ways through their implications for financial and economic stability, food security, and political security. The food crisis has added to general inflation and macroeconomic imbalances to which governments must respond with financial and monetary policies. At the same time, the financial crunch and the accompanying economic slowdown have pushed food prices to lower levels by decreasing demand for agricultural commodities for food, feed, and fuel.

The food price crisis has increased competition for land and water resources for agriculture, and declining capital for long-term investment due to the credit crunch has resulted in revaluation of natural resources. Farmland prices, for example, have been rising throughout the world. According to news reports in 2007 the farmland prices jumped by 16% in Brazil, by 31% in Poland, and by 15% in the Midwestern United States. Constraints in capital have also led to overexploitation and degradation of natural resources. In many countries, developed water sources are almost fully utilized, even as agricultural demand for water is expected to increase drastically in the future. Additionally, in the future, as climate change further increases climate variability, temperature, and the risk of droughts and floods, threats to agricultural productivity and production will rise (Von Braun 2008a).

5 Eco-terrorism

The global and national food systems are complex systems, which are typically characterized by non-linear and difficult-to-predict changes with sudden disruptions. Poor people are the least able to predict most of the risks in complex food systems and therefore, are also the most affected by their combined occurrences. Public policy must focus on preventing these risks (Von Braun 2009). On fringes of social and environmental movements there are always those who will resort to violence and sabotage namely, eco-terrorism. Overall, it would not be wrong to assume that these will somehow lead to an increase in international terrorism (Alpas and Kiymaz 2010).

Most research linking global environmental change (GEC) and food security focuses solely on agriculture: either the impact of climate change on agricultural production, or the impact of agriculture on the environment, e.g. on land use, greenhouse gas emissions, pollution and/or biodiversity. Important though food production is, there are also important new questions about the interactions between the governance of climate and food such as those associated with carbon trading and labelling, and the role of the private sector in carbon mitigation and in the management of food systems (Ericksen et al. 2009).

The risk of high and volatile food prices, which limit poor people's food consumption and other prerequisites of living, may increase in the future. According to IFPRI scenario analyses (Von Braun et al. 2008b), food prices are not likely to fall to their 2000–2003 levels in the next decade and price volatility is increasing. Financial and economic shocks, which lead to job losses, economic constraints and decreased demand for agricultural commodities, are likely to persist in some parts of the developing world. Thus, this fact may lead to a limited depressing effect on high food prices.

The impacts of climate change, such as droughts and floods result in a decrease in yields in developing countries may further exacerbate food insecurity. When the majority of the poor in the world are considered to be depended on agriculture as a source of food and income, the severeness of the climate change effects can be understood.

As an output of some of above-mentioned food system risks, more of the societal and political risks – such as food riots destabilization of governments and domestic and transborder conflicts – can be faced in the coming years. According to Von Braun (Von Braun 2009), in a successful development process, idiosyncratic risks gradually decline as public policies provide more insurance and safety-net coverage and the capacity of households to cope with adverse effects is strengthened. However, the food and financial crises pose global risks, which also may increase idiosyncratic risks. These two categories of risks are effective among the poor. The stylized pattern of risks in a framework of "decreasing–increasing" likelihood, and "small–large" severity of impact for the poor is given in Fig. 1. It suggests that most of the covariate risks are found in the "increasing" and "large" top

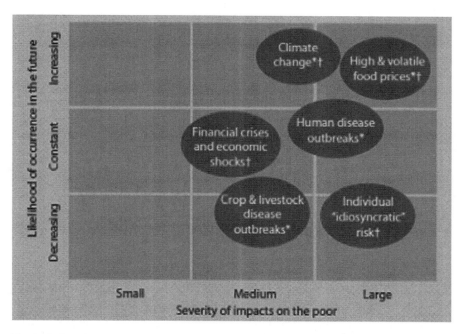

Fig. 1 Stylized framework for assessment of food security risks (Von Braun 2009). Notes: *=strong role of science and technology for mitigating severity and likelihood of risks; †=strong role of sound policies and institutions for mitigating severity and likelihood of risks

right corner of the risk space. As it is followed from the figure, the potential of food price volatility and climate change leading a rise in food insecurity among the poor groups is significant, and thus, can be expected to bring social disturbances and terrorism in the short to long term.

Ensuring food security while avoiding negative feedbacks to key ecosystem services places stiff demands upon the design and implementation of adaptation strategies and options (Ericksen et al. 2009). Food security is a vital and relevantly "hot" topic as all societies are crucially dependent upon the food supply; therefore, its disruption is an obvious prime target for terrorism (Alpas et al. 2010).

It was also recently highlighted that technical fixes alone will not solve the food security challenge, and the major environmental changes bringing additional threats to food security need a food system approach, not just a focus on agricultural practices. An equal emphasis was also given to the importance of global food trade in reducing vulnerability to environmental change, while also highlighting that non-production aspects of the food system (e.g. road and rail networks) can also be vulnerable to environmental change, on top of the obvious environmental stresses now facing agriculture in many parts of the world.

6 Biotechnology

A priori, biotechnology – one of many tools of agricultural research and development – could contribute to food security by helping to promote sustainable agriculture centred on smallholder farmers in developing countries. Yet, biotechnology is now a lightning rod for visceral debate, with opposing factions making strong claims of promise and peril (Serageldin 1999).

Innovations in agricultural biotechnology may be considered as an alternative to provide environmental benefits. This biotechnology revolution is very relevant to the problems of food security, poverty reduction, and environmental conservation in the developing world. But for many, it raises important questions relating to ethics, intellectual property rights, and bio-safety. There have been widespread protests against the spread of agro-biotechnology. Some of the concerns come from scientists who fear that "novel" products will destroy agricultural diversity, thus changing agricultural patterns into unrecognizable and uncontrollable forms. Many protests have been made by civil society institutions on ethical or ecological grounds. The dominance of a highly concentrated private sector has raised fears of a new phase of comparative disadvantage and increased dependency in the developing world (Serageldin 1999).

Looking ahead, there is no technological remedy for the global challenge of sustainable and secure food production. Biotechnology is considered to have a strong potential as one of the key technologies to offer solutions, accelerating our ability to develop varieties with characteristics of drought, heat and saline resistance, as well as resistance to pests and disease, although environmental and food safety conditions will need to be met. Genetic modification needs to be seen in the context of a large and diverse range of research areas relevant to food security. Global agriculture demands a diversity of approaches, specific to crops, localities, cultures and circumstances. Such diversity demands that the breadth of the relevant scientific enquiry is equally diverse and that science needs to be combined with social, economic and political perspectives (Royal Society 2009).

The international public research system has a critical role in ensuring that access to potential benefits of new technologies is guaranteed for poor people and environmental conservation. There must be recognition of the need for increased public involvement with biotechnology and for complementing private sector research, to ensure transparency and accountability and to promote a broad range of public goods research just as markets expand for results of private goods research. There is a need for win-win-win scenarios for all concerned actors, and for creative efforts to identify and put to work enabling mechanisms for the developing countries to benefit from the gene revolution. For the sake of today's poor, marginalized, and hungry people, and for future generations, this important challenge should not be forgotten (Serageldin 1999).

According to a recent study (Anonymous 2009a), biotechnology and advances in breeding have been helping agriculture to achieve higher yields and meet the needs of an expanding population with limited land and water resources. Production of primary food and feed crops, namely maize, wheat, rice and oilseeds, has increased by 21% since 1995, while the total cropland allocated to these crops has increased by only 2%. The productivity gains that accounted for most of the increased production of these crops are also important for global food security and land conservation. It is also stated in the same study that given the impacts of climate change on agricultural productivity and the part played by agriculture practices in global warming, agricultural techniques must play a substantial part in the fight against climate change. The three major contributions of green biotechnology to the mitigation of the impact of climate change are greenhouse gas reduction; crops adaptation; and protection and increase yield with less surface.

Green biotechnology offers solutions to decrease greenhouse gas (GHG) emissions and the tools to engineer crops using less energy through increase in yield. Green biotechnology crops can help to mitigate the climate change because they give farmers the opportunity to adopt a more sustainable farming system. Examples of the positive impact that could play the green biotechnology are less fuel consumption on farm; carbon sequestration; reduced fertilizer use (Barfoot and Brookes 2006).

As another example, the wide-scale use of glyphosate-based herbicides has reduced tillage considerably and helped conserve soil and stimulate soil microbial activity. One of the environmental benefits of low- or no-till agriculture is increased carbon sequestration in soil. Low-till agriculture is playing an important role in reducing the amount of carbon dioxide and methane released into the atmosphere. Although already impressive, future innovations in agricultural biotechnology have the potential to reduce greenhouse gas emissions on a much larger scale (Mehta 2003).

The main problem in maintaining further developments in biotechnology and propagating different types of novel environment-friendly plants in some of the countries is represented by the negative public opinion for the transgenic technology.

The genetically modified organisms (GMOs) and their use when necessary for the environment or because of environmental changes is often a sensitive issue in some of the countries or regions as a matter of oppositions for these kinds of seeds or agricultural products to be used in the field or for human nutrition.

7 Policy Choices

There is a great public concern about protecting the quality of the environment. This concern often shows itself in regulations that restrict agricultural production practices, such as application of agricultural chemicals. Harmful chemicals with long persistence in the environment and toxicity have to be banned with replacement of safer ones. Biological controls and integrated pest management are also important tools controlling pests, but may not be suffice. Biotechnology, which is criticized by environmental activists, is considered to have great potential not only to raise productivity of some products, but also to reduce the need for harmful chemicals.

The requirement of increasing agricultural production is likely to lead environmental danger. Agricultural production has to be increased by more than 20% in parallel with the estimated population growth in the coming 30 years by raising the area planted (FAO 2002). This would require massive destruction of forests, wildlife habitat and biodiversity and would reduce the carbon sequestration capacity of the forests. One of the acceptable alternatives is to increase productivity on the non-forest fertile soils and in animal production systems and the other is to reduce postharvest losses.

An environmental constraint on agricultural production is likely to be the adequacy of water availability. Water is priced to agriculture at zero in many countries; conversely, agriculture wastes water through animal wastes or chemicals and has no incentive to adopt water saving technologies. Greater incentives for agriculture to use water more efficiently is needed in this century. Otherwise water will become a severe constraint to world food supply, also due to climate change effects.

There are numerous powerful research tools available to develop environment friendly agricultural production technologies. At the same time, the public sector reduces its investment in agricultural research and private sector is far from compensating the reduced amount via new research projects. In several countries regulations restricts the ability of the private sector to apply some of powerful research tools like biotechnology. Many countries have to make the evaluation of the advantages and disadvantages of biotechnology, and where appropriate, have to provide intellectual property protection and decrease the cost of getting approval to make the genetically modified food available in the market. Overall there is substantial underinvestment in agricultural research to raise productivity on the fertile soils. Larger investments in agricultural research should increase production potential in every country while protecting the environment.

As a result of various initiatives that are being developed to reduce carbon emissions and environmental degradation (including carbon markets and bio energy developments) new demands are being placed on environmental goods and services and lands (including forests) are being assigned a monetary value. These initiatives may provide new opportunities for income generation and job creation; they are also likely to become more attractive to investors. This can result in insecure rights for the poor, including reduced access to the land or reduced ability to secure products. Therefore, these new opportunities, including bio energy, should ensure the participation and land rights of the people already present in the areas targeted for new initiatives (Anonymous 2009a).

To make food prices more stable and affordable for living population, governments may give their farmers a level playing field in which they pay/receive decent prices for the inputs and outputs. There is an important role for public investments in rural infrastructure, human capital and agricultural research as well as for providing a legislative system to support the efficient functioning of the market economy.

8 Food Security and Global Environmental Change: Vulnerabilities

Global environmental change (GEC), including land degradation, loss of biodiversity, changes in hydrology, and changes in climate patterns resulting from enhanced anthropogenic emission of greenhouse gas emissions, will have serious consequences for food security, particularly for more vulnerable groups. Growing demands for food in turn affect the global environment because the food system is a source of greenhouse gas emissions and nutrient loading, and it dominates the human use of land and water. The speed, scale and consequences of human-induced environmental change are beyond previous human experience, and thus science has a renewed responsibility to support policy formation with regard to food systems (Steffen et al. 2003; Carpenter et al. 2009).

A series of recent events have also stimulated broader interest in food security and food systems, most notably the 2008 news coverage of high food prices which were variously blamed on bio fuels, growing demand for meat and dairy products, commodity speculation and climate (Gregory and Ingram 2008). Other debates have arisen about the potential impacts of climate change on food availability and water as the projections of climate change become even more serious, around US/EU subsidies disadvantaging farmers in developing countries, and about the role of integrated policy in shaping food security in Europe and in other countries (Stiglitz 2006; Schmidhuber and Tubiello 2007). The price increases highlighted the connections between food systems in different places, e.g. drought in Australia and demand for meat in Asia, biofuel policy in the US and Latin America, and between the local food movement in Europe and export farmers in Africa.

Food systems can be described as comprising four sets of activities: (i) producing food; (ii) processing food; (iii) packaging and distributing food; and (iv) retailing and consuming food. These activities lead to a number of outcomes, many of which contribute to food security, and others which relate to environmental and other social welfare concerns. The challenges facing food systems will accelerate in the coming decades, as the demand for food will double within the next 25–50 years, primarily in developing countries, and with the WTO agriculture talks in disarray, options for reforming trade policy are still highly contentious (Von Braun 2008b). Food security and agricultural growth remain high on the science, policy and development agendas.

In this respect, vulnerability assessment could help address food supply-chain security by determining the selection of countermeasures and emergency responses. These assessments would also aid in the development of preparedness exercises and determine whether sufficient laboratory capabilities are in place. Such an approach would allow for the targeting of outreach to stakeholders through offering guidance as well as industry or regulator training. These measures accompanied by the development of lines of communication are critical in addressing the issue of food security (Alpas and Cirakoglu 2009).

9 Conclusion

Food security is directly affected by climate change effects that lead to concerns in rural livelihoods. In addition, economic or food supply crises led by causes other than climate change and/or policies causing high and volatile food prices are expected to occur in near future and to be factors of social unrests with severe impacts on the poor. For maintaining environmental security, agricultural policies which devastate the environment are to be replaced by rather controllable alternatives. Moreover, adapting to the additional threats to food security arising from major environmental changes requires an integrated food system approach, not just a focus on agricultural practices. Therefore it is very important to focus on existing research challenges for adapting food system to global environmental change (GEC).

The technological solutions alone are not sufficient for adaptation of food systems to GEC for increasing agricultural yields as bulk of the research is still focusing on agriculture not food systems. Also, tradeoffs across multiple scales among food system outcomes are a pervasive feature of globalized food systems and failure to recognize these often results in food insecurity, ecological degradation and loss of livelihoods (Sundkvist et al. 2005). On the other hand; resolving these tradeoffs, particularly across multiple levels of organization and decision making, is crucial to reduce the vulnerability of food systems to GEC (Ericksen 2008). In addition, there are some key underexplored areas within food systems as social and cultural values of food, nutritional implications of environmental change and ongoing changes in consumption patterns (Confalonieri et al. (2007). These are both sensitive to environmental change but it is also equally crucial to understand their implications for food security and developing adaptation strategies. The climate variability and climate extremes are likely to increase in near future, and changes in variability in climate are likely to have greater impacts on agricultural production than changes in mean climate alone (IPCC 2007). Therefore, much relevant information will need to come directly from the region(s) in question (Zurek and Henrichs 2007), including regional climate models as well as stakeholders.

Overall, vulnerability assessment could help address food supply-chain security by determining the selection and ranking of countermeasures and emergency responses as GEC will have serious consequences for food security, particularly for more vulnerable groups. In other words, either believed to be man-made or not, crises, incorporated in the food systems and accompanied by environmental degradations, have to be observed closely to foresee the right economic and social policies for preventing future cases of ecoterrorism and minimizing threats to food accessibility.

References

Alpas H, Ascl S, Koc A (2010) Food quality systems in Turkey: perspectives in terms of food defense. In: Koukouliou V, Ujevic M, Premstaller O (eds) Threats to food and water chain infrastructure, NATO Science for Peace and Security Series. Springer, The Netherlands, pp 25–48. ISBN 978-90-481-3544-8

Alpas H, Cirakoglu B (2009) Food chain security. Food Defense Workshop, CENS, Singapore

Alpas H, Kiymaz T (2010) Food security and ecoterrorism: recognizing vulnerabilities and protecting eco-systems. In: Liotta PH, Kepner W, Lancaster J et al (eds) Achieving environmental security: ecosystems services and human welfare. IOS Press, NATO Science for Peace and Security Series, Amsterdam, pp 239–249, (ISBN: 978-1-60750-578-5)

Anonymous (2009a) Bioenergy development: issues and impacts for poverty and natural resource management. World Bank Agriculture and Rural Development Notes, ISSUE 49 August 2009

Anonymous (2009b) Green biotechnology and climate change: www.europabio.org/positions/GBE/PP_090619_Climate_Change.pdf

Barfoot P, Brookes G (2006) GM crops: the first ten years – Global socio-economic and environmental impacts, ISAAA briefing No. 36 ISAAA: Ithaca, New York, ISBN: 1-892456-41-9

Carpenter SR, Mooney HA, Agard J et al (2009) Science for managing ecosystem services: beyond the millennium ecosystem assessment. PNAS 106:1305–1312

Confalonieri U, Menne B, Akhtar R et al (2007) Human health. In: Parry ML, Canziani OF, Palutikof JP et al (ed) Climate change 2007: impacts, adaptation and vulnerability. Contribution of Working Group II to the fourth assessment report of the Intergovernmental Panel on Climate Change, Cambridge University Press, Cambridge, pp 391–431

Ericksen PJ (2008) Conceptualizing food systems for global environmental change research. Global Environ Chang 18:234–245

Ericksen PJ, John SI, Ingram D et al (2009) Food security and global environmental change: emerging challenges. Environ Sci Policy 12:373–377

FAO (2002) World agriculture: towards 2015/30. FAO, Rome, 2002. http://www.fao.org/docrep/004/Y3557E/y3557e06.htm#e

FAO (2008) Climate change and food security: a framework document www.reliefweb.int/rw/lib.nsf/db900sid/PANA7KADCQ/$file/fao_may2008.pdf?openelement

Gregory PJ, Ingram JSI (2008) Climate change and the current "Food Crisis", CAB reviews: perspectives in agriculture, veterinary sciences. Nutr Nat Resour 3:1–10

Hoekman B, Martin W, Mattoo A (2009) Conclude Doha: it matters! Policy research working paper 5135, World Bank

IPCC (2007) Climate change 2007: summary for decision makers. Cambridge University Press, Cambridge

Mehta MD (2003) Climate change and biotechnology: moving toward a carbohydrate-based economy. Bull Sci Technol Soc 23:102–105

OECD (2008) Biofuel support policies – an economic assessment. Organisation for Economic Co-operation and Development, Paris

OECD (2009) Bioheat, biopower and biogas – developments and implications for agriculture. Organisation for Economic Co-operation and Development, Paris

Ortiz RM, Bellmann C, Hepburn J (2009) Agricultural subsidies in the WTO green box: ensuring coherence with sustainable development goals. Information note number 16, ICTSD: international centre for trade and sustainable development, Geneva, September 2009

Royal Society (2009) Reaping the benefits – science and the sustainable intensification of global agriculture. http://royalsociety.org/Reapingthebenefits/

Schmidhuber J, Tubiello FN (2007) Global food security under climate change. PNAS 104:19703–19708

Serageldin I (1999) Biotechnology and food security in the 21st century. Science 285:387–389

Steffen W, Sanderson A, Tyson PD et al (2003) Global change and the earth system: a planet under pressure. Springer-Verlag, Berlin/New York

Stiglitz J (2006) Making globalization work. Penguin Books, London

Sundkvist A, Milestad R, Jansson A (2005) On the importance of tightening feedback loops for sustainable development of food systems. Food Policy 30:224–239

Von Braun J (2008a) High food prices: the what, who and how of proposed policy actions. International Food Policy Research Institute, Washington, DC

Von Braun J (2008b) Food and financial crises: implications for agriculture and the poor. IFPRI, Washington, DC. www.ifpri.org

Von Braun J, Ahmed A, Asenso-Okyere K et al (2008a) High food prices: the what, who, and how of proposed policy actions, IFPRI Policy Brief 2. International Food Policy Research Institute, Washington, DC

Von Braun J, Fan S, Meinzen-Dick R et al (2008b) International agricultural research for food security. Poverty reduction and the environment – what to expect from scaling up CGIAR investments and "Best Bet" programs, IFPRI

Von Braun J (2009) Food-security risks must be comprehensibly addressed. IFPRI. http://www.ifpri.org

Zurek M, Henrichs T (2007) Linking scenarios across geographical scales in international environmental assessments. Technol Forecast Soc 74:1282–1295

Emergency Situations and Risk Management in Multilateral Environmental Agreements: A Comparative Analysis in the SEE

Gordana Petkovic

Abstract This paper addresses the issue of emergency response and risk management in the countries of South-Eastern Europe (SEE). A number of multilateral environmental agreements have been adopted by countries of SEE in order to better prevent disasters and reduce their consequences. However, those texts aim at strengthening international cooperation for emergency relief after environmental disasters (industrial or natural). In the light of considering the implementation of international obligations at the national level, the state of legislative and institutional framework in SEE is analysed and some specific actions are proposed.

Keywords Emergency situations • Risk management • Trans-boundary impact • Industrial accidents • Natural disasters • Prevention • Preparedness • Response • Notification systems

1 Introduction

Emergency situations generally constitute, both in terms of the areas where they occur and of the general consequences, one of the issues that affects, or that may affect relations among states and the basic values of mankind. Due to this fact and the universality of the environmental issues, this matter is becoming more and more important.

Adverse trans-boundary impacts from various human activities, especially from industrial accidents and natural disasters, have happened in the past and may happen in the future. However, during the second half of the twentieth century, several major chemical accidents occurred, causing extensive human suffering and widespread environmental damage (Seveso 1976, Bhopal 1984, Mexico City 1984, Basel 1986, Baia Mare and Baia Borsa 2000, floods, earthquakes, tsunami,

G. Petkovic (✉)
Ministry of Environment and Spatial Planning, Belgrade, Serbia
e-mail: Biser1956@yahoo.com

typhoons, desertification, and so on). As a result, extensive international efforts have been invested in order to improve accident prevention and preparedness for emergency response.

Such efforts were undertaken in the UN system through several UN programme activities, especially within humanitarian assistance (DHA), UN International Strategy for Disaster Reduction (ISDR), International Labour Office (ILO), United Nations Environmental Programme (UNEP APELL Programme-Awareness and Preparedness for Emergencies at Local Level), United Nations Economic Commission for Europe (UN/ECE), UNEP, UNDP and OESC (ENVSEC), WHO, IAEA, Council of Europe, Organization for Economic Co-operation and Development (OECD). Various soft-law instruments and multilateral agreements were adopted, dealing with the linkages between hazardous substances, industrial activities and consequences of climate change. A huge number of remarkable activities has been undertaken in the field of international management of industrial accidents and water resources. At the EU level, the so-called "Seveso Directive" was adopted by the European Community in 1982 and revised twice. Some of these activities are underway within the Danube Basin and Sava River Basin.

2 International Legislative Framework on Emergency Response and Risk Management

2.1 General Review

Multilateral Environmental Agreements (MEAs) play an important role in the protection and management of the global environment, especially those which were created within the UNECE region. The UNECE Multilateral Environmental Agreements are unique legal instruments which address the significant transboundary impact assessment from human activities in the Pan-European Region. The intention is to prevent environmental degradation from any activity which cause or may cause environmental damage in UNECE region, including the countries of South-Eastern Europe (SEE). Countries are aware of the importance that the implementation of these instruments may have in contributing to the environmental security in the region. The set of these legal instruments consists of five international environmental multilateral agreements within the UNECE region:

- Convention on the Trans-boundary Effects of Industrial Accidents (Helsinki Convention).
- Convention on Environmental Impact Assessment in a Trans-boundary Context (Espoo Convention).
- Convention on Long-range Trans-boundary Air Pollution and its eight Protocols.
- Convention on the Protection and Use of Trans-boundary Watercourses and International Lakes (UNECE Water Convention), Protocol on Water and Health and Protocol on Civil Liability.

- Convention on Access to Information, Public Participation in Decision-making and Access to Justice in Environmental Matters (Arhus Convention).

The Protocol on Civil Liability is a joint instrument to the Helsinki Convention and UNECE Water Convention but has not yet entered into force.

2.2 Linkages and Cross-Cutting Issues Between Multilateral Environmental Agreements

Risk is a basis of the States responsibility in case of trans-boundary effects of industrial accidents, while the direct link between risk and responsibility is an assessment of impacts on the environment. In the Helsinki Convention the risk appears as the central category and an assumption for the responsibility of contracting parties and other international law factors. This primarily arises, at the Convention level, from the following elements: the "hazardous activities" that are undertaken and which the Convention tends to regulate; the character of these activities and the possible impacts for the values that are protected. Indeed, the Convention envisages two degrees of possible effects of industrial accidents on the environment, including the effects of such accidents caused by natural disasters: the effects marked as "any direct and indirect or delayed adverse consequences" and the trans-boundary effects that may have "significant impact within the trans-boundary context."

Risk is an institute that, according to the Convention, includes and expands into a broad area of possible activities. This is best seen from the elements that constitute the "hazardous activities". The breadth of possible activities is additionally corroborated by the breadth of possible basic hazardous activity protagonists. For the purpose of the Convention, it is anybody in charge of an activity (monitoring, planning, implementation or prevention of activity), while from the point of view of general determination it is the contracting party ("the party of origin" or "the affected party").

In order to better understand risk content, it is worth mentioning the possible answer that the Convention provides to the question on what might be a risky event. It is generally defined by the elements that make is as: an uncontrolled event, relating to dangerous matters that may be in an installations or partly in transport; and those that in certain conditions may cause a "hazardous activity". This leads us to conclude that the emergence of a "hazardous activity" that may cause transboudary effects is, in fact, the central category the Convention was entered into for.

In addition, the Convention tries to regulate more thoroughly the issues relating to the definition of "hazardous activities" and its general spirit is characterised by the strive for the undertaking of preventive measures. Thus, the estimate of "hazardous activities" overlaps in some way with the risk assessment, and/or assessment of environmental impacts.

When a hazardous activity is subject to an environmental impact assessment in accordance with the Espoo Convention and that assessment includes an evaluation of the trans-boundary effects of industrial accidents from the hazardous activity which is performed in conformity with the terms of this Convention, the final decision taken for the purposes of the Espoo Convention shall fulfil the relevant requirements of Helsinki Convention. In this way, the responsibility of States for possible industrial accidents is directly linked with an adequate risk assessment and arises from the same. Here, the risk appears of the State's responsibility for providing that activities within their jurisdiction or control do not affect significantly the environment of other states or areas beyond their national jurisdictions. Also, the obligations arising from the Convention for the contracting party require that the States undertake the measures relating to the establishment of legal regulations, administrative, financial and technical measures for their prevention and operational preparedness to respond adequately to industrial accidents.

In accordance with the Convention, the Parties have the following obligations:

- Identification and notification of hazardous activities according to Annex I to the Convention (substance and quantity criteria) – similar to Annex I of the "Seveso II" Directive.
- Emergency preparedness and response:
  - establish and maintain adequate preparedness and response measures, including on-site and off-site contingency plans.
- Industrial accident notification:
  - notify the affected country/countries in the event of a major industrial accident with trans-boundary effects (at national, regional and local levels);
  - establish and operate notification systems at the national, regional and local levels;
  - designate a national point of contact – operational 24 hours a day.

In setting-up of prevention and preparedness measures, the public has the right to access to information and access to administrative and judicial proceedings in line with the Aarhus Convention.

The Conference of the Parties (COP) accepted the UNECE Industrial Accident Notification System as an early warning tool for notifying a major accident at national level. The UNECE IAN System comprises: a network of national points of contact; a procedure and set of notification forms to be used; a system of reports, including an early warning report, an information report and an assistance request report. The COP decided to establish and operate the UNECE Past Industrial Accident Reporting System within the EU-MARS (Major Accident Reporting System) – joint work with the EU Major Accident Hazards Bureau.

Bilateral cooperation include: mutual assistance in case of a major accident within the UNECE Industrial Accident Notification System, scientific and technological cooperation, exchange of information (exchange information, experience and technology related to prevention, preparedness and response to industrial accidents) and information to and participation of the public (inform the potentially affected public in the neighbouring country and provide opportunity to participate in relevant procedures).

The importance of the Convention, related to the purpose and subject of its regulation, lies in the fact that it requires the States, including the countries of South-Eastern Europe (SEE), to improve their legislation, especially in line with the EU requirements, strenghten the institutional capacities and establish efficient early warning system.

The Water Convention is intended to strengthen national measures for prevention, control and reduction of pollution, monitoring research and development, consultations, warning and alarm systems, mutual assistance, institutional arrangements, and the exchange and protection of information, as well as public access to information. This Convention served as a basis for the elaboration of such subregional agreements such as Danube Protection Convention. The Convention specifies the details for the actions to be taken: prevention, control and reduction of pollution, including trans-boundary impact, protection of the water resources, emission limitation, setting up emission inventories and action programmes, monitoring, reporting, consultations, communications, warning and alarm systems, mutual assistance, institutional arrangements, and the exchange and protection of information, as well as public access to information.

The aim of the Convention on Long-range Trans-boundary Air Pollution and its eight Protocols is to limit and, as far as possible, gradually reduce and prevent air pollution including long-range trans-boundary air pollution.

The Convention on Climate change together with the Kyoto Protocol imposes emission limits and further reduction of greenhouse gases emission, which results in necessary actions taken up in many sectors of the economy, including energy, transport, agriculture, industry, construction and forestry.

2.3 The Integrated Approach Related to Water and Emergency Response in SEE

At the First Meeting of the Parties to the Framework Agreement on the Sava River Basin (FASRB) which was held on 1 June 2007, the Parties underlined the significance of cooperation and development of integrated approach in the Sava River Basin (SRB). Multilateral environmental agreements (Helsinki Convention, Water Convention, Arhus Convention) have not been ratified yet by all Parties to the FASRB, although they have incorporated liabilities, arising from these conventions and relevant EU directives, into their national legislation. All these instruments stipulate the cooperation on international, regional, and respectively bilateral levels. Taking into account the basic principles and requirements emanating from international legal instruments and regulations of the European Union in this area, the adoption of the protocols to the FASRB would be aimed at harmonization of the water management regime in the Sava river basin considering the features and specificity on a regional and respectively the national level.

In the Declaration from the 2nd Meeting the Parties of FASRB the need for enhancing trans-boundary cooperation on water resource management was pointed out, through sharing of best practices, joint development on climate change scenarios, and coordinated and cooperative approaches at the regional level, as well as for building adaptive capacities of institutions in each Party to manage emerging challenges, particularly those related to climate change. In addition, the Parties recall that climate change is likely to affect the water management activities in the Sava river basin, and supports further investigation of its possible impacts, and development of integrated approach, that involves mitigation and adaptation actions, as well as closely related projects. The Parties emphasized the importance of strengthening the ability to prevent, control the effects of, and respond effectively to emergency situations arising from human-caused and natural disasters. The Parties expressed awareness of the necessity to promote the development and operation of compatible and efficient regional cooperative mechanisms for disaster control and preparedness, including early warning systems for environmental emergencies recognizing experiences at other regions levels and national capacities.

The Protocol on Trans-boundary Impacts in the Sava river basin would be aimed at regulating the procedure for the issuance of water law acts for installations and activities that might have trans-boundary impact to the unity of the water regime in accordance with the mechanisms of the Espoo Convention, the Water Convention and the EU requirements.

The Protocol on Emergency Situations, pursuant the relevant international legal instruments of UNECE, would have the goal to establish a risk as central category and assumption of derivation of liabilities of the Parties, respectively to this protocol and other relevant protocols to the FASRB. This protocol would stipulate two degrees of possible impact of the accident, including industrial accidents and natural disasters to waters, water regime and aquatic eco-systems:

- effects marked as "direct or indirect release of substances, which might have detrimental effects to human health or quality of aquatic ecosystems, into waters"; and
- trans-boundary effects involving "significant impacts" in trans-boundary context.

This especially relates to countries' liabilities in the field of trans-boundary effects of accidents, and the direct link between the risk and liability is the impact assessment in the Sava river basin.

A Protocol on Floods would be aimed at the regulation of issues regarding sustainable protection against floods in the Sava river basin caused by elementary forces and artificial impacts with the objective to prevent or limit hazards and reduce or eliminate the negative effects of floods. The Parties to the protocol would establish cooperation based on the Action program for sustainable protection against floods in the Danube River Basin, as well as in the EC Directive on the Assessment and Management of Flood Risks (2007/60). Each country should prepare the preliminary flood risk assessment for its share of the SRB, as well as the Flood Mapping Inventory. The protocol also would stipulate the preparation of the Flood Risk Management Plan, along with the establishment of the informing, forecasting, warning and alarming system.

A Protocol on Prevention of Water Pollution caused by navigation has been signed at the 2nd Meeting of the Parties of FASRB, Belgrade, 1 June 2009. The aim of this Protocol is the prevention, control and reduction of pollution originating from vessels, the establishment of technical requirements for equipment of ports, the development of the best available techniques, informing, response measures and water quality monitoring. Water quality monitoring requires a network of national institutions for monitoring and inspection. Trans-boundary cooperation should include actions to prevent pollution from vessels by developing a joint action programme. Informing the competent authorities of downstream countries on location, quantity and polluting substances assumes use of all means of communication including radio network for early warning of pollution. Efficient reactions in case of accident, which means release, spillage and disposal of substances from vessels, that has or may have trans-boundary impacts, requires the set-up of an institutional framework, meaning the establishment of a joint body for determination of reasons and facts relating to the accident, its impact on the environment and possible pollution on the section of the waterway.

The main liabilities originating from the protocols principally refer to:

- Establishment of identification, consultation and mutual information in regard to all activities, including industrial activities that may cause trans-boundary impacts effects for which the impacts assessment is undertaken (for existing and planned activities).
- Setting-up of the risk assessment and risk management measures (hazardous activities, including natural disasters and industrial accidents).
- Introduction of long-term measures for risk reduction, based on the development and implementation of new safe technologies and overall development planning, in accordance with overall environmental capacities.
- Establishment of water quality monitoring on national level.
- Setting-up of the appropriate system for informing, forecasting, warning and alarming, realization of cooperation and mutual assistance in case of announcement or actual accident that could cause trans-boundary effects (from enterprises to municipal, regional or state level).
- Cooperation in the development, research and exchange of knowledge, information and technology in the field of risk management, as well as in the development and implementation of new technologies.

2.4 The Impact of European Union Legislation on SEE Countries

The European Union (formerly European Community) is a Party to the relevant UNECE international agreements. Moreover, the EC Directive 96/82 on the Control of Major Accidents Hazards Involving Dangerous Substances (Seveso Directive), along with the specific directives on IPPC, flood protection, air protection and waste management has a broad impact on the legislative and institutional framework of the member states as well as of the SEE countries, with regard to:

- laying down of a system of prevention of major accidents;
- prohibition of the activity of an installation that do not comply with the requirements of the Directive;
- specification of the installations or group of installations with higher potential of risk of major-accidents as a consequence of the mutual position of these installations;
- implementation of a system of elaboration and approval of the required safety documentation;
- preparation of the external emergency plans;
- participation of the public in the approval of dangerous activities and information of the public, as well as on how to behave in case of occurrence of an accident;
- implementation of a system of controls of installations with a risk of major accidents.

Specific requirements of the Directive are related to the following issues:

- Prevention:
  - specification of enterprises, that could not cause a major accident hazard;
  - laying down criteria for the assessment of potential danger in installations with higher risk of accident;
  - providing a guidance for solving the problems related to the storage and use of dangerous substances and supplying information included in the notifications, in safety reports and emergency plans;
  - including criteria for land use planning concerning the dangerous substances;
  - allocating sites for installations, potentially carrying a danger of chemical substances.

- Preparedness and response:
  - preparation and implementation of a preparatory plan for the state administration and enterprises;
  - introduction of procedures for acquiring, exchange and dissemination of information on accidents;
  - issuance of provisions for reporting on dangerous substances in specified enterprises;
  - providing for operators to report to the administrative authority the location of the installation and the amount of the dangerous substances;
  - establishment and implementation of a system of control and carrying out regular controls;
  - providing that given institutions prepare external emergency plans, based on the information, presented by the operators;
  - evaluation of individual plans of new enterprises with dangerous chemical substances;
  - providing that the internal and external emergency plans, are regularly evaluated, revised, and if necessary, renewed and that the public is fully informed.

- Notification system, competent authorities and focal point:
  - implementation of a system of notification;

- appointment of the competentnt authority and if necessary establishment a body for technical assistance to the administrative authorities;
- appointment of an institution responsible for elaboration of external emergency plans.

• Public participation through in-advance established procedures:
  - in the administrative procedures for land use decision-making related to the installations with dangerous chemical substances;
  - through giving opinion to the problems under discussion and with prospect that these points of view will be considered in the assessment of notifications, safety reports and emergency plans;
  - in the procedure of adoption of safety reports, list of dangerous substances and information on the danger of major accidents;
  - in the procedure of preparation of the emergency plan.

• Exchange of information:
  - implementation of a procedure for providing exchange of information between the enterprise and the public in case a major accident or serious impacts is likely to happen;
  - providing information to the Member State, that might be affected by the impacts of industrial accident with trans-boundary impacts;
  - establishment of information systems that would provide acquiring of relevant data on installations and accidents (enterprises that cannot cause major accident, major accidents and results of their analysis; experience gained in preventing major accidents; transposition and implementation of the Directive and so on).

3 A Comparative Analysis in SEE

The Helsinki Convention on the Trans-boundary Effects of Industrial Accidents has been ratified by 36 States and the European Community (now European Union) and entered into force on 19 April 2000. The Parties to the Convention from SEE are EU member countries (Bulgaria, Greece, Romania and Slovenia) and EU non-member countries (Albania, Croatia, Serbia). Bosnia and Herzegovina, Macedonia, Montenegro and Turkey have not yet become the parties to this Convention.

3.1 Competent Authorities

All the countries have appointed competent authorities. In Slovenia, the Ministry of Environment and the Ministry of Defence are the responsible authorities, while in Croatia, Montenegro and Romania the Ministry of the Environment and the Ministry of Interior were chosen. The Competences are split between the Ministry of Environment, the Ministry of Defence and the Ministry of Interior in Serbia.

The Ministry of Environment and the Civil Defence Service are the competent authorities for the implementation of the Convention in Bulgaria. In addition, specific institutions have been established, such as the State National Situation Centre, to serve as information centre which collect, analyse and classify information on emergency situations. Moreover, the State Agency for Civil Protection is contact point for notification of the population in Bulgaria. The General Inspectorate for Emergency Situations has been established by the Ministry of Interior in Romania. In Turkey the competences are split between the Ministry of Environment, the Ministry of Labour and Social Security and The Center of Poison Information.

SEE countries that were not yet parties had identified authorities responsible for implementation of the Convention. Some of them have some difficulties, such as lack of an appropriate administrative structure (Bosnia and Herzegovina) or an insufficient cooperation and coordination between the authorities involved (industry, environmental protection, mineral and energy resources, transport) (Albania).

3.2 Implementation Issues

Most of the Parties have established legislation on emergency situations and risk management. However, the description of legislation provided in most reports sent to the Secretariat of Helsinki Convention were quite general and only a few countries provided references to the specific articles of the Convention (Serbia). Legislation seems to be fully in place and in force in most of the Parties in Western and Central Europe and also in Bulgaria and Romania. Judging from the reports of the countries of SEE, the extent to which appropriate legislation is in place in these countries still differs significantly. Mostly, disasters are regulated by environmental and/or defence, civil protection legislation (Albania, Bulgaria, Croatia, Romania, Serbia, Slovenia). In Serbia it seems that legislation is to a significant degree adequate, while in the Macedonia legislation seems to be well under way. It seems that the main problem is the practical implementation of this legislation.

SEE countries analysed their legal and institutional frameworks with regards to shortcomings and identified future actions to improve the situation. These working areas, together with cross-cutting areas, were also considered in the draft strategic approach for the Assistance Programme's implementation phase. Bulgaria, Romania and Serbia had been accepted to the implementation phase of the Assistance Programme. The working areas were: identification of hazardous activities, notification of hazardous activities, prevention, preparedness, response and public participation. The cross-cutting areas were legislation and institutional frameworks.

Macedonia has benefited from awareness-raising mission UN/ECE (2007), which had been allowed to draw up an action plan for implementing basic tasks under the Convention. Bosnia and Herzegovina was in an initial phase of implementation of the Convention. Legislation on prevention, preparedness and response is under development. Albania has been facing difficulties in enforcement of legislation. Towards implementation of the Convention, Albania has developed concrete steps such as identification (adjournment) of Competent Authority, identification and information on hazardous activities (especially in cooperation with neighboring countries) and establishment of notification system.

3.3 Identification of Hazardous Activities

Some SEE countries provided lists of hazardous activities that seemed to be based on different criteria than Annex I of the Convention. Among the EU members, the SEE countries, namely Slovenia and Romania, have identified and notified hazardous activities, while Greece did not. Bulgaria is still faced with the question of notification. Among the EU non-members, Albania, Croatia and Turkey have not yet identified and notified hazardous activities. Bosnia and Herzegovina and Serbia have a problem with regards to the notification. Albania and Serbia have faced the problem of insufficient expertise in identifying hazardous activities, location criteria and risk analysis. Macedonia is in its initial phase of identification of hazardous activities.

Moreover, among SEE countries, only Slovenia has developed bilateral cooperation in transboudary context of industrial accidents. The Cooperation Agreement on the Forecast, Prevention and Mitigation of Natural and Technological Disasters was signed on 18 July 1992 in Vienna between Austria, Croatia, Hungary, Italy, Poland and Slovenia. The Convention for Cooperation in the field of Environmental Protection was concluded between Bulgaria and Romania (UN 1992). The cooperation covers exchange of information about disasters and accidents form nuclear plants, chemical installation and other activities which can affect the territory of another country through emission of hazardous substances in the atmosphere, water or soil. An intergovernmental committee has been set up with the representatives of both countries.

3.4 Prevention

Some countries have undertaken measures referring to Seveso II Directive implementation or to legislation rather than to practices description of preventive measures including: verification of safety documentation, facility inspections, issuance guidelines for operators safety management system and risk management. The SEE countries still have problems related to insufficient cooperation and coordination between authorities responsible for ensuring safety at hazardous activities (Serbia), insufficient know-how on risk assessment methodologies, risk management and safety standards (Albania, Bulgaria, Romania) and dialogue between authorities and operators (Serbia).

3.5 Notification and Mutual Assistance

Presently there are 43 members of the Industrial Accidents Notification network. Some 24 countries have established regional/local notification systems. Serbia and Macedonia have established institutions responsible for industrial accidents notification. However, they were not yet officially designated as points of contact under the IAN System. In addition, it is recognized insufficient implementation of the IAN System, especially information exchange with neighbouring countries (Serbia) and inadequate emergency response equipment (Albania, Serbia).

3.6 Emergency Preparedness

The level of detailed information varies and only a few SEE countries have data regarding revision of the plans or testing procedures (Slovenia, Bulgaria). The level of cross border cooperation in practice is uncertain. Serbia and Albania have recognized insufficient contingency planning. In addition, there is insufficient compatibility between contingency plans in a trans-boundary context (Serbia). That is the reason for drills and exercises to test contingency plans, particularly in a transboundary context (Serbia).

3.7 Participation of the Public

Most of the Parties have formally implemented the provisions regarding public participation, but information on how the legislation is applied is still not available. Many countries only refer to implementation of the Aarhus Convention. Some countries still do not give potentially affected public in neighbouring countries access to such participation or equal access to judicial and administrative procedures.

3.8 Decision-Making

Very general information exists regarding the land-use planning and decision-making on the location of hazardous activities except in Slovenia. Most of the countries handle these issues through their land use planning systems. Usually, the SEE countries referred to EIA procedures in national and trans-boundary contexts. Some countries apply risk assessments in this regard, but no country referred to particular methodologies or specific risk acceptance or decision criteria.

4 Conclusion

Environmental security, especially with regard to the human dimension, is a major issue in emergency situations. Industrial accidents, but especially natural disasters, can result in environmental degradation and major displacements of populations who are victims or at risk. The aim of the multilateral environmental agreements is to highlight risk prevention through a proactive approach. Countries should engage to undertake various measures, such as modification or adoption of plans (e.g. location of installations, coastal zones and rivers).

In order to apply Helsinki Convention, particularly in order to establish an effective emergency response system, the countries need to strengthen capacities for:

- identification of hazardous activities and safety measures;
- establishment and maintenance of emergency plans;
- notification systems and operational response capabilities;
- horizontal and vertical coordination between authorities as well as cooperation/communication between the authorities, the industry and the local communities.

The implemetation of the actual know-how represents a priority and, in this sense, assistance capacity building activities may be identified in:

- training and education of inspectors on the identification of the causes of chemical accidents and on best practice for preventing accidents;
- assessment of the quality of safety reports and safety management systems;
- definition of best practice on coordinating the activities of different inspectorates;
- improvement of international cooperation.

Practical implementation of legislation and procedures need to be strengthened through the Assistance Programme and its Strategic Approach.

Within the FASRB activities is important to encourage efforts in drafting and ratification of protocols to it. The exchange of information on undertaking the activities that are subject to environmental impact assessment in a trans-boundary context would ensure integrity of the water regime in the Sava River Basin. Protection against hazardous impacts of waters involves the realization of a series of facilities and other activities for the protection of people, natural and man-made goods from floods and all types of water erosion. Liabilities set-up in the Protocol would impact the obligations of the Parties to undertake measures at national level with regard to the development of legal regulations, the establishment of administrative, financial and technical measures for prevention of accidents, as well as the bringing to a level of operative readiness to respond to accident adequately, including floods protection and prevention of water pollution caused by navigation. Principally, this refers to obligations of the competent authorities and holders of industrial and other activities, for the purpose of establishment of prevention measures, preparedness measures and response in case of emergency. The implementation of FASRB protocols could have direct and long-term impact to different branches of economy, which use water resources or make impact to them in their activities (agriculture, industry, energy, inland transport, fishery, tourism). Economic implications relate to the more intensive possibility to participate to regional cooperation and projects implemented in respect of the control and reduction of transboundary impacts, in the field of monitoring and assessment, development of a system of warning and alarming based on the elements of AEWS system developed by the International Commission for the Protection of Danube River (ICPDR).

In order to reach the Millennium Development Goals, countries should take actions to develop and implement prevention and risk reduction measures integrated into all parts of sustainable development strategies. Climate change should be a key component of this activity. Moreover, it would be necessary to improve and support environmental emergencies response, both at a strategic and operational level. Response to the climate threat should be supported with concrete measures with the aim to reduce global greenhouse gas emissions, minimize risks for environmental disasters and provide a more extensive adaptive response.

References

Directive 96/82/EC on the control of major accidents hazards involving dangerous substances
Draft protocol on trans-boundary impacts to the FASRB, 2009
Draft protocol on floods to the FASRB, 2009
Draft protocol on emergency situations to the FASRB, 2009
ECE (2007). Sixth ministerial conference "Environment for Europe" Belgrade 10/12/October 2007. Implementation of UNECE multilateral environmental agreements (ECE/BELGRADE CONF/2007/12), 18. July 2007
ECE (2008). United Nations Economic Commission for Europe, www.unece.org, Fourth report on the Convention's implementation (2006–2007), ECE/CP.TEIA/2008/3 of 1 September 2008
Protocol on prevention of water pollution caused by navigation to the FASRB, 2009
UN (1991). Convention on environmental impact assessment in a trans-boundary context, United Nations, 1991
UN (1992a). The convention on the trans-boundary effects of industrial accidents, United Nations, 1992
UN (1992b). Convention on the protection and use of trans-boundary watercourses and international lakes, United Nations, 1992

Large Scale Infrastructural Projects in South Eastern Europe and Their Impact on Political Relations, Economic Development and Environmental Security

Dimitar Pekhlivanov

Abstract The economic backwardness of the region of South Eastern Europe is obvious – it lags behind from the most developed European countries in many terms like income per capita, level of corruption, effectiveness of public sector and so on. In the first part of this contribution, an analysis of the main weaknesses of the public sector in the region is provided, putting in the centre the low confidence among the stakeholders in all the society. A special part is devoted to the *low interpenetration* among the different SEE countries, especially in border areas, confirming the role of EU as an "anchor" for the development of the region.

The second part is dedicated to the enumeration of some of the most interesting large-scale infrastructural projects going through the region, like gas and oil pipelines, bridges and others with the main data, level of preparation and main obstacles and problems. Concrete examples and analysis of the main expected benefits from the realization of the given (and other) large-scale infrastructural projects is given. A reference is made to the environmental risks and problems which may emerge during the realization of similar projects, even with *"finest preliminary tuning."*

Keywords SEE • Economic backwardness • Public sector • EU • Infrastructural projects • Regional development • Environmental risk

1 Introduction

The problem with the political, economic and social development of the South-East European countries is one of the main problems of the overall development of the European continent. For many decades and by different reasons this development has been contemplated as one of the key prerequisite of the stability on the European continent.

D. Pekhlivanov (✉)
Institute of Public Administration, Sofia, Bulgaria
e-mail: pechdimi@yahoo.com

When we look back at the economic history of the world in the last 50 years, there are several cases when a country lagging behind in economic terms has converted itself into a highly developed economy with a strong private sector and effective public policies. The cases of South Chorea, Chile, Singapore, Taiwan have one important thing in common and this is the presence of a strong and mature partner, playing the role of an *"anchor"* with inflows of investments, know-how and all kinds of experiences. This role was played by powers like Japan for the Asian continent and the USA for many American countries. Similarly, such role was played by the European Economic Community (then EC, then EU), for countries like Greece, Ireland and Portugal. The EU will play this role also for the Balkan countries, no matter of their formal EU membership (Romania and Bulgaria), clear and relatively near membership (Croatia) and more distant and somehow vague membership (Macedonia and Turkey). In any case, the EU and these countries have to find and use different kind of instruments for stimulate and boost their overall development. This tools have to increase, firstly, the political weight of states from the region, stimulate the private entrepreneurship, import and really implement its best practices, invest in know-how and last, but not least, make stronger and more effective one of the most sensitive political problems in the region – the public policies and generally the public sector.

These instruments for accelerating the economy of a country or a region could be different, but among them, beyond any doubt, are large-scale infrastructural projects in different sector like transport (airports and highways), water management (dams or waste water treatment plants) and energy projects (gas and oil pipelines). The list could be prolonged almost indefinitely and examples for the highly positive influence of similar projects could be found everywhere in the world. Obviously, the importance and impact of these projects, especially trans-regional ones, go far beyond their practical use, engaging, firstly, the highest policy makers, attracting powerful foreign investments, know-how, trade and travel flows, stimulating the local economies in all sectors and, the most important – increasing the future political "price" of the respective countries. This is valid especially for the South-East European states, where the big infrastructural projects reflect, like in a mirror, the most "hot" topics in the region: low effectiveness of the public sector, substantial delays in implementation, high levels of corruption, lack of political will due to historical prejudices, necessity to conform with local lobbing and other interests or just fear to be engaged more than necessary with the neighbours. In that sense the international infrastructural projects, routing through the South-Eastern Europe, or let call them, *"transbalkan"* projects, should play an important role in the overall development of the region, discovering, from one side, the mentioned weaknesses and helping enormously, from the other side, to overcome them. Projects like the South Stream and Nabucco gas pipelines, the Second Danube bridge, the new highways, the Burgas – Vlorë, Burgas – Alexandroupolis and Constanta – Trieste oil-pipelines in any case will influence the development of the region for many years and decades ahead.

The idea of the present work is to contemplate more closely the interdependence between the construction, development and the future full running of these projects and the overall development of South-Eastern Europe; in other words, their impact on the region. To be more precise, since this impact for us is out of doubt and is mainly positive, the main focus should be on the extent and scope of this, generally accepted, positive influence. Using mathematical quantities, the present paper will try to find and to determine the *"percentage of usefulness,"* hidden behind the realization of these projects. Moreover, this influence will not be the same in all areas of social and economic life, in any case more for one and less for others – we shall try to find the impact as precisely as possible. Besides that, these infrastructural projects should have, as said, a generally positive impact both for the region and for the continent, but their impact on several sectors, like the environment, for example, is ambivalent. Moreover, the benefits from these projects could be roughly divided into benefits during the construction (public procurements, local sub-contractors, etc.) and those to be assessed in the longer term (transit tax earnings, increased political "price" of the country, new investments etc.).

When trying to make an academic research on similar topics and especially on South East Europe, there are several difficulties which should not be underestimated. First of all, all these projects are time-consuming and their full completion is quite remote in the future (especially given the experience in implementing large-scale infrastructural projects in the region), in particular with regard to the calculation of the potential benefits after the project's completion. In such case, it is really difficult to estimate the real impact in the future, since no real figures are available. The experience from similar projects in other parts of the world could be a suitable reference, but in that case the special features of the region are not enough taken into account. Secondly, given the fact that many of these projects are still *"captives"* of the political considerations and even bargaining, the detailed academic researches on that topic are really scarce and the main analysis, conclusions and the possible recommendations have to be made on the basis of documents, media articles and Internet based information, which are not always reliable sources. Of course, this is also a challenge, because it gives more freedom for different kind of interpretations and discoveries, but this scarcity of solid academic background on the topic is quite visible and should be taken into account.

2 Economic Backwardness, Weakness of the Public Sector and Political Immaturity of South-Eastern Europe. How to Overcome Them?

2.1 Economic Backwardness and the Weakness of the Public Sector in the Region

The economic backwardness of the region is one of its main features and one of the most discussed topics in all levels of the communities and decision making establishment. The economic problems of the region are visible especially at the *"micro"* levels, because some of the countries in the region have quite satisfactory macroeconomic indicators. However, looking at different classifications and indexes, made by different institutions and concerning the quality of life, the income per capita and especially the effectiveness and in fact the weakness of the public sector, these countries are placed well behind the Central and Western European countries and, most importantly, close the one to another.[1] Of course, the economic problems of the region cannot be regarded as something absolute, but predominantly in relative terms – to the rest of the European continent. In comparison with the average world levels of income per capita, life expectancy and quality, the countries from the region are relatively stable, much above the average world values. One of the other features of the region is the differences inside the region in economic, political and social terms. The most advanced countries in the region, Slovenia and Greece, are much ahead from the least developed states or regions like Albania. The differences between Slovenia and Albania, for example, are comparable to those of two countries from two extreme edges (Nord-South or West-East) of the European continent.

The reasons behind this political and economic variety of the regional map and the backwardness of the region are as complex as the differences themselves. They are hidden firstly in the historical heritage of the region, bearing not only the stamp of the centuries long Ottoman empire domination, but also of the Soviet domination in some countries for the last half century. Also a lack of historical chance could be found like the displacement of the main trade and commercial flows from Mediterranean to the Atlantic during the Middle Age. Here we can add also the permanent contradictions between the big European states and the Ottoman Empire, which led to dismemberment of some countries in the region and to somehow *"administrative"* borders, not always conformable to the real and historically established ethnic borders.

Generally speaking, the reasons for the backwardness of the region are various and the short list presented here is far from being exhaustive. The much more important for the present work is the results of the region's backwardness in terms of situation in the public sector, level of mutual trade and commercial flows. If we look, for example, at the data with the incomes per capita in the world,[2] including also the countries in the region, we can draw some interesting conclusions.

In the first place, well ahead are the *"flagmen"* of the region, Greece and Slovenia, which even could be to some extent excluded from the different analysis treating the region, since they differ substantially from the general trend. In the second place, most of the countries are forming a compact group, ranging from place 49 (Romania) to 65 (Macedonia) and three of them following one another, which is very symptomatic in a variety of 166 countries. The differences among the most of the countries in this group are practically almost negligible and not only in terms of incomes per capita. Respectively, three countries are *"detached"* from this

---

[1] The most interesting indexes as income per capita, corruption index, life satisfaction are given in the text as references and not as tables due to the restricted space.
[2] http://siteresources.worldbank.org/DATASTATISTICS/.

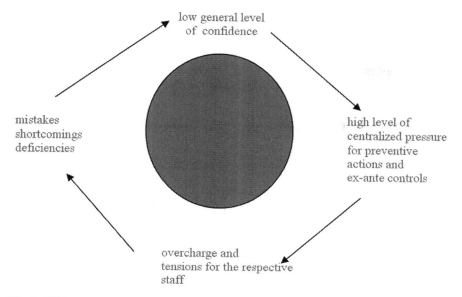

Fig. 1 "Vicious circle" of the public sector in the states from the region

group (Greece, Slovenia and Croatia) ahead and two other (Bosnia and Herzegovina and Albania) are lagging behind.

Similar are the data concerning the Global Corruption Index,[3] presented in the data of the Transparency International. One of the differences in the above table is the presence of Greece in the *"compact group"* of countries from the region, but nevertheless, the above range confirms once again the general trends for the region – many common problems, similar level of economic and social development and substantial lagging behind the most developed European and American countries.

Furthermore, the circle below represents a sort of analysis of the main weaknesses of the public sector in the region. The analysis does not pretend for a universality for all the countries in the region, but pretends to present a more or less general model. It begins and ends with the general *low confidence* among the main stakeholders in the society – state, ministries, agencies, municipalities, beneficiaries, companies, etc (Fig. 1).

The analysis concerns mainly the management of public resources in terms of realization and completion of infrastructural projects, in other words the capacity of absorbing public financial resources, both national or European. As already mentioned, the starting and the end point of this circle is *the low confidence*[4] in the

---

[3] http://www.transparency.org/policy_research/surveys_indeces/cpi/2009.

[4] The low general confidence in the states from the region is confirmed by many researches and other sources, therefore is chosen to be the starting point of the present *"vicious circle"* analysis. No to forget that the confidence is directly related to the levels of corruption.

society as a whole. The feeling of the low confidence is somehow increased by the very low level of life satisfaction, also visible in the global Satisfaction of life Index,[5] where the results are even worse – the people from countries like Bulgaria, Romania, Albania, Bosnia and Turkey are among the most *"unhappy"* in the world.

Hence, the *low level of confidence* evokes and requires high pressure for centralization and different, sometimes absolutely superfluous checks, additional controls, developing of analysis and strategies (*right part of the circle*). The tight controls and centralization aim at the increase of the confidence among the different stakeholders and the guarantee of high quality of disbursement process, which is quite understandable in countries with relatively high levels of corruption and deriving of public finances for private purposes. But if we add the non-motivated and very often non-competent administration, this high pressure leads to an overcharge of the respective staff and higher shortcomings of resources in terms of time, worse coordination, constant changes etc. (*low part of the circle*). This overcharge and overpressure lead to mistakes and shortcomings in the essential part of the disbursement and implementation process (*left part of the circle*). Hence, the mistakes, shortcomings and irregularities lead to a new decrease of confidence among the main stakeholders and to a *new starting point of the mentioned vicious circle*.

It is quite difficult to analyze where and how the circle could be lacerated. For us, the analysis of the critical issues in the process of public expenditures is weak or missing and does not lead to real effective measures for decreasing the bureaucracy and the extremely overcharged and complicated procedures. In other words, this analysis has to be stronger and more effective and lead to a real discharge and simplification of disbursement process.

2.2 Low Level of Interpenetration in the Region

One of the other very visible features so typical for the region is the very low level of mutual *interpenetration*. This phenomenon is described very clearly in all researches, dealing with the general level of the development of the region, especially the different kind of transport infrastructure.[6] This term of the *low interpenetration*, quite visible especially in the border regions, means mainly weak and somehow sporadic connections among the countries in the region. More concrete, the interconnections among these countries in all kind of infrastructure like transport (land and air connections), energy infrastructure (gas and oil pipelines) and many others are weak and far from the levels of "real" European states. The *low interpenetration* among the states in the region, of course, has its many reasons and explanations. Firstly, taken into account the level of infrastructure in the individual states, we hardly can expect that the interconnections among them could be better.

---

[5] http://www.le.ac.uk/users/aw57/world/sample.html.

[6] A quite interesting research in that sense is "Business in the Balkans. The case for cross border cooperation" Liz Barrett, Centre for European Reform, July 2002.

Indeed, the level of transport and other infrastructure in the individual countries from the region is far below those in the Central – and Western European countries, whereas it is easy explicable why the regional interconnections are also very weak.

The low interoperability between the transport systems of the different countries and the low number of cross-border checking are quite visible.[7]

Similarly, the air connections between the different capitals are not so *"abundant,"* since the national air carriers, being state property in most of the countries, serve also as a flagman of the national identity.[8] As a general tendency, the level of these connections is relatively weak, compared with similar regions in other European regions – Northern Europe or Benelux. There are, for example, missing connections between several capitals like Sofia – Belgrade or Sofia – Zagreb, Zagreb – Belgrade, Bucarest – Tirana and many others. Vienna is the capital serving the most often as a hub for the air connections between the capitals in the region. Very often a fly on a given itinerary in the region is really a peculiar experience.

The situation with the trade regimes is very similar, although in the past decade the trade connections among the SEE countries from the region were substantially improved in terms of establishing free trade with each other. Practically now there is not a country without any agreement with its neighbours, with the only exception of Kosovo. Nevertheless, this does not help substantially for the regional trade, since no State from the region has the main trade partners among its immediate neighbours.[9] In fact, the EU is by far the main trade partners for all the countries in the region.

The *low interpenetration* in the region is much far interesting if we try to answer the question *"why?"* As the first reason one can, as mentioned, indicate the general relatively low level of economic development in the region, which presupposes a low level of transport, trade and other relations. The second very interesting and important reason is *"locked"* in the past, whatever period we can take as a reference. For example, the strategic world confrontation between the two military blocks, NATO and Warsaw pact, has practically frozen every serious attempt to facilitate the contacts between their different members. This is very visible in the case of Bulgaria from one side and Turkey and Greece, from the other. For decades there was only one cross-border check between Bulgaria and Turkey – for almost 300 km of border[10] – and the situation was similar between Bulgaria and Greece and Bulgaria and Yugoslavia. Sometimes, however, this past played a positive role like in ex-Yugoslavia, since the transport infrastructure in the federation has been planned and realized on federal level, taken into account the national particularities and there were practically no cross-border checks. The past and the different

---

[7] For example, the number of cross-border check points between the individual countries can be checked very easily and it is much times less than on the borders in Central and Western Europe.

[8] A detailed information on that could be found in the Internet-sites of the respective national air companies.

[9] As said in the Study on Danube Bridge (see reference list) "the trade relations are inexcusably weak for neighboring countries."

[10] The second has been opened in 2005 only.

disputes all over the history played, from other side, a very important role in the development of the relations between Greece and Turkey. To a very great extent, this problems hidden in the history are not yet overcome and in any case will influence the relations between the two countries for many years ahead. The historical reminiscences in the region are something very typical, influencing very seriously the regional development. In any case, no one serious researcher can omit them. They influence the thinking of the highest political decision-makers, as well as the mass moods which, in turn, influence the decision-makers. From that point of view, serious and for the moment difficult to overcome are the problems between Bulgaria and Macedonia, Turkey and Greece, Macedonia and Greece and many others. The "historical" moods in the region can be expressed by the title of the very famous book of Jane Austen "Pride and Prejudice," in the sense that in many cases they are not even trying to establish closer relations with the neighbours or, if establishing them, they are not striving to their development. The emotional degrees of the political relations in the region, kept by the historical feelings, determine to a great extent the relations in the region. This is an additional evidence of the political immaturity of the region – the history is not left to the historians as it has to be – the history really tries to determine the political decisions, the history influences to a substantial extent also the media and their editorial politics. Generally speaking, the history, its reminiscences and its role for the development of the region deserves a special research, far beyond the scope of the present paper. Here we have just argued how big the role of the history is and how much it influences the region.

Looking at the more recent past, we can find additional arguments of the non-development of common infrastructure – on first place, of course, the wars in ex-Yugoslavia in the first half of the nineties of twentieth century. The reasons of the wars, the damages caused on the infrastructure, on the interstate relations and even on the human psychic, is very complex to be explained here, although it is clear that this wars played their roles of additionally dividing the region. In fact, on the one side, the most advanced countries in the former Yugoslavia, namely Croatia and Slovenia, tried to place themselves among the countries of Central Europe, striving to fall in another very different political orbit. On the other side, less developed ex-Yugoslav countries, such as Macedonia and Bosnia and Herzegovina and to some extent Serbia, have dropped behind, loosing the necessary impetus for catching up the momentum of the European integration process in the region. As an additional, but not decisive remark, given the weak development of the region and the political and historical prejudices, on several places the interconnections between the countries are impeded by the natural forms of the landscape like big rivers (Bulgaria and Romania), mountains (Greece and Albania) or other natural hurdles, which, of course, play a more substantial role in case of historical prejudices and lack of political will.

Undoubtedly, the other main reason for the low level of mutual penetration, together with the historical prejudices, is the fear of isolation. There are many examples in the last decades of how the states from the SEE region have rejected every serious attempt for unifying themselves, fearing that this unification could have a negative impact on the perspective of the European integration. Therefore, it could be argued that the main political slogan of the region should be "Not together, but together in Europe!."

2.3 European Union – Main "Anchor" for the Region

Treating the general question of the present work, we have to clarify what we are expecting exactly as an answer when we are looking at the large-scale infrastructural projects, named *transregional* ones. In other words, are these projects a real instrument for changing the region? Are there some opportunities for the region to help itself somehow *"from inside,"* by using entirely internal resources? Looking at all the aspects of the development of the region, including the improvement of the ecological security, the general economic development and, last but not least, political maturity, one conclusion comes as an answer – the European Union and forces equal to it, as an external force are *"doomed"* to be an *"anchor"* for the region, as mentioned quite extensively in the introduction. The region could hardly find internal reserves for deciding its very specific, we should say, endemic problems. These projects, being part from the political agenda and investment package of the big external forces like USA, Russia and EU, can exercise both pressure and control over the region and there is not another way. For example, it is very interesting to see how the different disputes in the region have been influenced and finally decided by the EU in the last years. Such examples can be given with the border dispute between Slovenia and Croatia which was on the edge to suspend the accession negotiations of Croatia. Similarly, the real, at first glance purely administrative accession to EU has enormously pushed ahead the relations between Bulgaria and Romania and Bulgaria and Greece.

3 Large-Scale Infrastructural Projects – A Tool for Pushing the Regional Development

3.1 The Projects "Nabucco," "South Stream," "Second Danube Bridge" and "Burgas-Vlorë" – "Flagmen" of the Regional Development

Nabucco is beyond any doubt one of the most ambitious European energy projects in the last decades. Actually the European states receive gas supply from different sources, whereas Russia is by far the main supplier for the continent. The energy policy and the gas and oil supply, however, have been transformed in the last decade in the flagman of the Russian foreign policy, trying to impose the Russian influence by the "energy diplomacy" (Fig. 2).

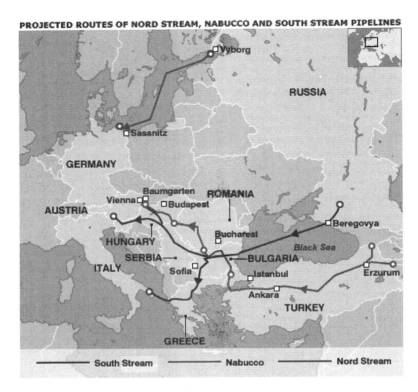

Fig. 2 North Stream, Nabucco and South Stream pipelines

This Russian *"energy diplomacy"* has evoked a reaction from the EU side, aiming at decreasing the dependence from Russia in terms of gas import. This reaction accelerated lastly after the Russian–Ukrainian crisis from the winter of 2009. Besides the Nabucco project, the European Commission has developed a new proposal for Regulation on the security of gas supply, which does not address the question of where the natural gas should come from, but it is a step towards a more comprehensive, a more centralized and a more effective European energy policy.

We have to say right from the beginning that Nabucco is both an economic project, but also a *"policy pipeline,"*[11] a result from the energy political game between EU and Russia. In second place, Nabucco is a reply to the decreasing opportunities for insuring gas for the European states in the middle and far future. According to the data of IEA – the World Energy 2008, the demand of gas in EU will approximately in the years after 2020 increase more and more, attaining the impressive quantity of 600 bcm/y.[12] To be very correct, Nabucco does not presuppose a fully independence from Russia, which is practically impossible – the project gas-pipeline capacity of 31 bcm/y is no more that 10–15% of the overall gas import to the EU.

---

[11] Interviews with Joschka Fischer, "Capital News paper", Sofia November 2009.

[12] According to "South stream" and "Nabucco" presentations.

The overall length of the gas-pipeline is supposed to be 3,300 km from the Eastern border of Turkey to the grid system in Baumgarten near Vienna. The international agreement for the construction of Nabucco was signed in August 2009 in Ankara – Turkey.[13] The construction phase of the project is to begin in 2011.

Nevertheless, the future of Nabucco as a gas-pipeline is quite uncertain due to the political problems among the EU member states and the availability of sufficient and reliable sources of gas to reach the planned capacity of 31 bcm/y. The gas fields in Azerbaidjan are far from being sufficient for that purpose, the other planned sources like Uzbekistan and Turkmenistan have already signed contacts for selling gas quantities to Russia by long-term contracts and on reasonable prices, moreover Turkey and Azerbaidjan have trade disputes on the gas-price. The filling of Nabucco with gas from Iran is theoretically possible, but very difficult in practice because of technical reasons, not to forget the political problems with USA.[14]

South Stream is the next large-scale infrastructural project of European scale, passing through several countries in the region and to a very substantial extent influencing the region. The project foresees the construction of a gas pipeline through the Black sea, whereas the offshore part consists of 900 km pipeline. The two gas-pipelines, South Stream and Nabucco, pursue the same goals: filling the increasing gap between the demand and the own production in the European continent and the diversifying of the gas routes to the European consumers (Fig. 3).

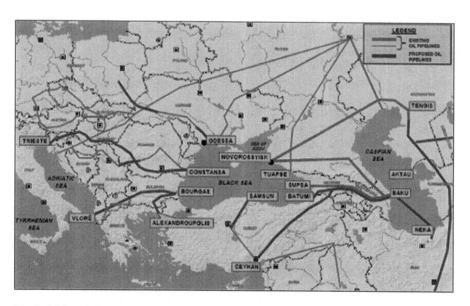

Fig. 3 Main existing and proposed oil-pipeline in Eastern Europe

---

[13] Official site of Nabucco www.nabucco-pipeline.com.
[14] Димчо Петракиев, "Набуко – шансове и препятствия, Център за анализ в енергийната сфера, София," (Dimcho Petrakiev, "Nabucco – Chances and obstacles", Centre for analysis in energy field, Sofia 2009).

South Stream and the already existing Blue Stream gas-pipeline in the Eastern part of Black Sea represent a part of the Russian strategy for increasing the Russian influence through energy supplies. Besides that, this projects serve as a tool for pressure on countries like Bielorussia and Ukraine, since they will deprive this countries of quantities of transit gas and forwarding it directly to EU countries. The very discussed rivalry between the both gas-pipeline (Nabucco and South Stream) lies mainly in the political importance of the projects and, more especially, in the political symbolism of both projects: – the first, simply said, tries to decrease the dependence from Russia, and the second, tries to do the contrary.

However, although being a "*private,*" according to different sources,[15] project of "Gazprom" and the Italian ENI, we can say that South Stream is much more effective in terms of already concluded agreements with the transit countries. Moreover, many of the EU countries prefer sure supply now (even though Russian) than uncertain supplies in the future. The feasibility study of South Stream is expected in 2010, as well as the contract with the local South Stream companies.

The Second Danube Bridge is the only among the enumerated projects that is in implementation (constructive) phase – although the delays registered till now reflect the already mentioned problems in public sector (e.g. Introduction). This new way on Danube river, foreseeing new routes and rails, will substantially shorten the way to Central and Middle Europe, passing through EU members only. With the expected entry of Romania and Bulgaria into Shengen zone, the way (and time) for lorries and all kind of transport to Europe will be really facilitated. The project has the strong support of the European Union and its financial institutions, being part of the Corridor N4. Besides that, the project will substantially improve the situation in poor and bleak regions of Romania and Bulgaria.

One of the most serious planned projects in the region, unfortunately also not started till now, is the "*transbalkan*" Burgas-Vlorë oil-pipeline or the so called AMBO (Albania, Macedonia and Bulgaria Oil) oil-pipeline. The history of the project starts in the beginning of the 90s by initiative of the newly created Macedonian state willing to receive oil also from Burgas harbor (not just from those in Salonika), but the idea gradually evaluated from "*macedonian*" to "*transbalkan*" with the corresponding conception, size and investments (in terms of distance it has been "*prolonged*" to Vlorë and in terms of general conception it changed from helping Macedonia to "*survive*" into the idea to sell cheaper and "faster" oil westwards). The project has a strong American support, confirmed by many political agreements related to the signature of the tripartite Memorandum and the official Convention between Albania, Macedonia and Bulgaria.

AMBO is really a project, which deserves the attention of biggest oil-industry investing companies in the world.[16] Firstly, it substantially shortens the way from the Black Sea, Caspian and Central Asian oil fields to the Western and North American markets and it escapes away from the Straits by easing so the traffic there, mitigating the risks

---

[15] Interview with Joschka Fischer, "Capital Newspaper."
[16] First Oil-forum "Energy route to Europe&USA Through the Balkans, Belgrade, September 2009.

of crashes and oil-floods. The *"competing"* oil-pipeline Burgas – Alexandroupolis also escapes the Straits, but finally it seems to be more expensive in terms of final accounts, being far away from the western markets and having predominantly local importance. Even if not enumerating all the advantages of Burgas-Vlorë pipeline, one thing deserves attention – the fact that it is a "pure" Balkan project – it starts and finishes in the countries from the region. Most of the other considered projects in the present paper begin in the countries of the region, but their final destination is Central or Western Europe (Nabucco or South Stream). In this case the building of the pipeline will really test the capabilities of the transit countries to cope with large-scale projects and more especially the capabilities of their public sectors to control, analyze, monitor and finally complete similar undertakings.

3.2 Main Expected Benefits from the Realization of Large-Scale Infrastructural Projects

What should be the direct answers of the many posed question of the present work? To say right from the beginning, counting all the direct and indirect benefits for the region is really impossible. Our look is more *"strategic"* than *"immediate,"* firstly because some of the immediate benefits like transit fees and taxes are quite obvious (although not going directly into the national budgets) and secondly, because according to our opinion, the region needs a deep and universal development, based on the principles of modern democracies in political terms and on those of the sustainable development in social and environmental terms.

Looking for an overall change of the *"climate"* in South Eastern Europe and being entirely in the spirit of the present paper, we could argue that the main benefit from the realization of the listed projects is political. An elementary truth in the theory of the international relations is that a country with pipelines on its territory *"increases"* its political price. Besides that, looking carefully on the Nabucco's map, most of its transit countries are already (although relatively new) EU members. – In this sense the gas pipeline will *"add value"* to the already existing political weight of those countries as EU members. Moreover, the importance of the gas-pipeline, in terms of increasing the inter-dependence and political responsibility of those countries to manage well the project, will *"discipline"* the decision-makers charged with the project's management.

The investments expected in the construction phase of all these projects are huge – just Nabucco and Burgas – Vlorë account at approximately 10 bln. EURO's.[17] Even distributed among the different states on pipeline's track, it represents a serious investment for the respective economies, given also their relative weakness. One major part of this investments relates to the raw materials – the expected quantity of steel for Nabucco, for example, is about 2 mln. tones of steel,[18] which will be

---

[17] Some of the information concerning Nabucco are still confidential (feasibility study, for example). Nevertheless, the official estimations are enough to calculate the impact.

[18] Official site of Nabucco www.nabucco-pipeline.com.

provided mainly by local contractors and/or by subcontractors. According to the experts in the field of realizing similar projects,[19] the constructive phase of a pipeline (usually oil or gas) attracts at least 15–20 or more big contractors, not to count the subcontractors. The outsourcing of all the activities (including also on the spot accommodation and catering for hundreds of people) related to the construction will attract a huge numbers of additional contractors. Just as an example – a lorry, specialized in transporting the different segments of the pipeline (pipes) can transport no more than 3–4 parts, approximately 10–15 m each or 40–50 m altogether. Calculated to a distance of 900 km of pipes (if we take as an example Burgas – Vlorë oil-pipeline) it is even difficult to imagine how many lorries (and transport companies) are needed to submit all the segments of the pipeline.

Or let take another very concrete example, which *"stretches"* far into the future: – the communication channels (cables), which are laid on together with the pipes for the future communication needs of the pipeline, could be made larger and the *"superfluous"* capacity can be sold beforehand by a contract to the telecommunication companies on both sides of the borders. This can be an additional tool for increasing the communication among the countries from the region and overcoming the *low interpenetration* among them.

Another very important strategic benefit from the realization of these projects is the development of new or secondary industries like, for example, the development of infrastructure related to the construction of UGS (Under Gas Storage) facilities, the development of Electric Power industry and the Gas chemistry, as well as the enlarged utilization of gas as a vehicle fuel. All these "developments" will multiply the expected impacts on the local environment.

Confining entirely to the spirit of the present work and looking for strategic benefits for the region rather than for short term palliative measures, one of the main long term benefits can be found in the improvement of the public procurement procedures, which are the main source for corruption and bad practices. The tighter international control on the public procurement and the strongly respected procedures during the implementation period could be able to improve substantially the overall mentality and thinking on the respective officials and members of administration.

One of the most important long term benefits, according to our opinion, is the development of the border regions and all kind of contacts among them[20] – at first place the improvement of the *local access* by roads and rails, opening of new cross – border check points, development of local business. In that sense, one of the requirements during the construction of pipelines is the full accessibility to every track's point, which, quite understandable, will require the construction or the rehabilitation of new roads.

---

[19] Interview with Roumen Nikolov, official representative of AMBO Corporation for Bulgaria, Sofia, January 2010.

[20] Study on Inter-regional Measures for the Development of Middle Danube Region, Project financed under PHARE CBC Bulgaria-Romania program, Louis Berger SAS, Euroconsultants S.A..

As said before, as far as the political aspects are concerned, these projects will increase the "political price" of the region and of the individual countries. Moreover, several public relations activities can be developed with regard to the construction of similar projects – the activities, related to the construction can be brilliantly fit into a well developed and balanced national Communication strategy.

3.3 Main Environmental Problems and Risks Related to Large-Scale Infrastructural Projects

The obligatory objectivity of every academic research requires also to enumerate the most important environmental risks and problems for every similar large-scale infrastructural project. Due to the worsened respect of environmental norms in the countries from the region, the usual embarrassments of similar constructions for the population could transform themselves into a very detrimental event for the nature in the WA (Working Area) of the respective projects. The problems increase from the fact that the Ecological Compliances are good "on paper" and change completely during the real implementation; moreover a lot of unexpected circumstances can appear. At first place, the track of the pipelines should strictly go round the sites from Natura 2000 network in the different countries. In the case of Burgas-Vlorë oil-pipeline, for example, there are several sites, just in Bulgaria – the Reserve "Atanasovo" lake, one National park ("Rila") and one Natural park (Vitosha), as well as some natural sites in Macedonia. Nevertheless, even under the conditions of *"finest preliminary tuning"* of the construction works, some of the possible environmental and other risks during the implementation of the projects can be summarized as follows:

- Surface disturbance due to WA (Working Area) grading different from original profile.
- Dust production due to earthworks.
- Hydrologic interactions, due to trenching and water stream and channel crossing. Pipeline crossing in main water streams and channels will require increased depth of cover (minimum depth of 3 m) and reinforced coating. Pipeline crossing of secondary channels will be performed through horizontal drilling, with a minimum allowable depth of 1 m under the channel, and with mechanical protection up to 1.5 m depth.
- Road transportation (traffic of heavy-duty machinery and pipe) and temporary road diversions, road downtime due to pipeline sections installed across communication ways.
- Waste production and dumping of excess material from trench excavation.
- Interference with human activities, due to accidental breakage or damage of underground services (electrical or telephone cables or other underground facilities).
- Interference with human activities, due to temporary interventions for keeping all sewers, drainage and other natural water courses open and functional.

- Higher noise emissions.
- Hydrologic interactions, due to the creation of a physical barrier and/or preferential drainage route.[21]

There is not a full guarantee against all these and other risks. According to the official information available till now, all the necessary measures for the respect of environmental norms will be taken. Obviously, much more efforts will be needed for respecting these norms also during the implementation phase. In any case, all the available reports and analysis till now indicate that the benefits, also in environmental terms, will prevail over the environmental risks. Looking back to the beginning, we could pretend that the usefulness of the similar projects will be much more than 50%.

4 Conclusion

As indicated many times in the introduction and the paragraphs of the present work, the question about the large-scale or *"transregional"* or *"transbalkan"* infrastructural projects in the region of South Eastern Europe is very complex, with numerous and mutually interlacing aspects. More in detail, there are many problems – every single aspect from the realization of this projects seems to be problematic. The most important political aspects are still caught in the struggle between different present and future interests – among the EU states, between EU and Russia, between the Middle East and the European countries. The same is valid for the economic and financial parameters of these projects – to which extent are they remunerative and in what period their reimbursement could be expected, whether they are bankable or not. Moreover, some of the projects pursue purely political goals which, in a certain moment, can prevail over the economic benefits or vice versa support a project, which is not enough beneficial. The same is valid for the ecological dimension, since the ecological standards in the SEE region are generally existing just *"on paper"* and are not respected in many cases. All these aspects are more complicated by the normally low level of development of the region and by the non respect of the principle of sustainable development. It is also absolutely possible that some of the considered projects will be finally not carried out.

Nevertheless, we can expect that such projects could make a breakthrough in the *"witched"* borders of the region, by linking this region fast and easily to the Western Europe – the *"common dream"* of the local political decision-makers. What about *"our dream"* – this is to contribute to the change of the political and human climate in the region, bringing it much closer to the modern world.

---

[21] Many of the enumerated problems are listed also in AMBO Feasibility Study and Study on Interregional Measures (Danube Bridge).

References

Business in the Balkans – the case for cross-border cooperation, Liz Barrett, July 2002. Centre for European Reform – http://www.cer.org.uk./
Integrating the Balkans in the European Union: functional borders and sustainable security – IBEU work package: regional cooperation
NATO economic colloqium (June and July 1993) Brussels upgrading the Balkans regional infrastructure, by Qerim Qerimi; Bruno S. Sergi, SEER – South-East Europe review for labour and social affairs, issue 1/2006
Study on inter-regional measures for the development of middle Danube region (Danube bridge). Project financed under PHARE CBC Bulgaria-Romania program, Louis Berger SAS, Euroconsultants, SA
1st Oil Forum – Belgrade. 09.2009, Presentation
AMBO feasibility study, 2000 – AMBO corporation, financed by the US Trade and Development Agency (TDA)
"Burgas-Alexanrdoupolis Oil Pipeline Agreement" between Bulgaria, Greece and Russian Federation
Convention on "Burgas – Vlorë Oil-pipeline" between Albania, Macedonia and Bulgaria – text final
Study on inter-regional measures for the development of middle Danube region, Project financed under PHARE CBC Bulgaria-Romania program, Louis Berger SAS, Euroconsultants, SA
Pre-investment study of Danube bridge, Institute for transport and communications SocioEco Danube bridge V3 Final, December
Interview with Joschka Fischer. Political consultant of Nabucco pipeline, "Capital Newspaper", Sofia, 9 November 2009
Interview with Steven Man. State department energy coordinator. Conference "Natural Gas for Europe – Partnership and Secutiry", Sofia, 24–25 April 2009 – http://www.energysummit2009.bg/en/
The South Stream Pipeline and the Environmental Factor – www.Balkananalysis.com (7/18/2009)
Димчо Петракиев, "Набуко – шансове и препятствия", Център за анализи в енергийната сфера, София, 2009, (Dimcho Petrakiev, "Nabucco – chances and obstacles", Centre for analysis in energy field, Sofia 2009)

Internet Sources

http://siteresources.worldbank.org/DATASTATISTICS/
http://www.transparency.org/policy_research/surveys_indeces/cpi/2009
http://www.le.ac.uk/users/aw57/world/sample.html
http://www.tarom.ro/
http://www.croiatiaairlines.hr/
http://bhairlines.ba/
http://www.jat.com/
http://www.air.bg/
http://www-wds.worldbank.org/
http://www.gmfus.org/
http://south-stream.info/
http://www.nabucco-pipeline.com;/
http://www.mfa.bg/
http://www.energysummit2009.bg/en/

Environmental Impact Assessment in a Trans-boundary Context in the SEE Countries

Nataša Đereg

Abstract Environmental Impact Assessment (EIA) as a tool used to identify the environmental, social and economic impacts prior to decision-making in transboundary projects has been introduced in countries of South Eastern Europe (SEE) relatively recently.

This paper gives an overview of the existing legal regimes and obligations regarding EIA procedures in the SEE region taken in the process of EU integration. It presents the main elements of the Espoo Convention and the bilateral/multilateral agreements under the Convention aimed at enabling its effective application, as well as a short analysis of the implementation of the Espoo Convention in the SEE region so far. Through this paper, we discuss the new multilateral agreement under the Espoo Convention signed among countries of the SEE – the Bucharest Agreement, its purpose and applicability, as well as the identified constrains. We argue that mechanisms for public participation in the Bucharest Agreement need to be clearly established and that the special regime proposed for the joint cross border projects is jeopardizing the principles of the Espoo Convention and other UNECE Conventions. Considering that the SEE region have yet to declare Natura 2000 areas, and that many already protected areas are near the borders, we recommend broadening the criteria on significance under the Bucharest Agreement to take into account the biodiversity richness of the region.

Keywords South Eastern Europe • Environmental Impact Assessment • EU integration • Espoo Convention • Bucharest Agreement • Public participation • Trans-boundary projects

---

N. Đereg (✉)
Center for Ecology and Sustainable Development,
Belgrade and Subotica Korzo 15/13, 24000, Subotica, Serbia
e-mail: djnatasa@yahoo.com

1 Introduction

Environmental impact assessment (EIA) procedure has been introduced in South Eastern Europe (SEE) relatively recently either through specific legislation or as part of general environmental protection laws. EIA procedure aims at evaluating the likely impact on the environment, human health and safety of the proposed project/ activity before issuing a building or an operation permit.

Most of the SEE countries have adopted an EIA legislation due to the obligations taken with regard to the EU integration processes and the obligations under the Treaty establishing the Energy Community.[1] Proper implementation of the EIA procedures is also requested by donors and investors, particularly by International Financial Institutions such as the World Bank and the European Bank for Reconstruction and Development (EBRD) REC (2003).

SEE Countries that are members of the European Union have aligned their legislation fully with the Community *acquis* at the moment of accession or even before (Greece 1 January 1981, Bulgaria and Romania 1 January 2007).

Other countries of South Eastern Europe Albania, Bosnia and Herzegovina, Croatia, Kosovo (UN administered territory), The former Yugoslav Republic of Macedonia, Montenegro, and Serbia are at various stage of the accession process that implies implementing the Community *acquis* in the field of environment, and consequently the European EIA legislation: Directive EC 85/337 of 27 June 1985 on the Assessment of the Effects of Certain Public and Private Projects on the Environment and amendments with Directives EC 97/11 and 2003/35. While most countries have already transposed EC Directive 85/337 into their national legislation with some minor discrepancies, full implementation requires considerable improvement, especially regarding provisions on public participation and the consultation process, the trans-boundary dimension of the EIA procedure, the overall duration and inconsistencies in EIA procedural steps (Energy Community Secretariat 2008) (Table 1).

Table 1 Review of the EIA legislation and Aarhus Convention in non EU part of the SEE region (European Commission 2009)

Country	Directive 85/337/EEC	Aarhus Convention, ratification
Serbia	Transposed	31 Jul 2009
Albania	Transposed	27 Jun 2001
Croatia	Transposed	27 Mar 2007
Montenegro	Almost fully transposed	2 Nov 2009
FRY Macedonia	Almost fully transposed	22 Jul 1999
Bosnia and Herzegovina	Almost fully transposed into Entities' laws	1 Oct 2008
UNMIK Kosovo	Not in line	Kosovo is not a member of the UN and therefore has not signed the Aarhus Convention

---

[1] Signed in October 2005 in Athens and entered into force on 1 July 2006, between EU and SEE countries (Albania, Bosnia and Herzegovina, Bulgaria, Croatia, the former Yugoslav Republic of Macedonia, Romania, Serbia, Montenegro and the Special Representative of the Secretary General on behalf of the United Nations Interim Mission in Kosovo).

2 Espoo Convention

2.1 Introductory Notes

The main principle of international environmental law applied to the control of environmental pollutants as stated under section 21 of the declaration of the United Nations Conference on the Human Environment in Stockholm in 1972, is related to the responsibility of each State to ensure that activities within their jurisdiction or control do not cause damage to the environment of other States or of areas beyond the national limits.

Following this approach, a regional convention under the auspices of the United Nations Economic Commission for Europe (UNECE) was adopted in Espoo in 1991 – the Convention on Environmental Impact Assessment (EIA) in a Transboundary Context (The Espoo Convention).

As of December 2009, there were 44 parties to the Convention (43 UNECE countries and the European Community).

The Convention has been amended twice. The first amendment to the Convention was adopted in 2001, and once in force it will enable other UN Member States not members of the UNECE to join the Convention. The second one was adopted in 2004, and once in force it will allow, as appropriate, affected Parties to participate in scoping, require reviews of compliance, revise the Appendix I (list of activities) and make other minor changes. The SEA Protocol of the Convention (Kiev 2003) – Protocol on Strategic Environmental Assessment, once in force, will broadened the Convention's area and will require its Parties to evaluate the environmental consequences of their official draft plans and programmes in the areas of agriculture, forestry, fisheries, energy, industry including mining, transport, regional development, waste management, water management, telecommunications, tourism, town and country planning or land use.

Countries of the SEE region have all ratified the Espoo Convention, as it is showed in Table 2.

Table 2 Status of ratification of the Espoo Convention, its amendments and protocol in the SEE region

Country	Espoo Convention	First amendment	Second amendment	Protocol on SEA
Albania	4 Oct 1991	12 May 2006	12 May 2006	2 Dec 2005
Bosnia and Herzegovina	14 Dec 2009	–	–	–
Bulgaria	12 May 1995	25 Jan 2007	25 Jan 2007	25 Jan 2007
Croatia	8 Jul 199 6	11 Feb 2009	11 Feb 2009	6 Oct 2009
Greece	24 Feb 1998	–	–	–
Montenegro	9 Jul 2009	9 Jul 2009	9 Jul 2009	2 Nov 2009
Romania	29 Mar 2001	16 Nov 2006	–	–
Serbia	18 Dec 2007	–	–	–
FYR Macedonia	31 Aug 1999	–	–	–

The Espoo Convention is designed to promote environmentally sound and sustainable development while also enhancing international cooperation in assessing environmental impacts, particularly of trans-boundary nature- across national borders. The Espoo Convention primarily establishes a trans-boundary EIA regime for a planned project in one country likely to have a significant impact on the environment of another country.

The Espoo Convention describes principles, provisions and procedures to be followed and establishes the list of activities (Appendix I) for which the international EIA procedure is obligatory if there is significant environmental impact that may extend across the borders. The concerned Parties should discuss and determine whether an activity from the list may have significant impacts. Also, the Parties themselves should decide about the application of the Convention on all other activities- not listed in Appendix I, if there are special circumstances, by applying Appendix III of the Convention that contains general criteria to assist in the determination of the environmental significance. As the Convention relies on national EIA procedure, there are useful criteria and thresholds to be used from international, national and regional environmental programmes and treaties. There are two other UNECE Conventions and their respective lists in the Annexes which should also be taken into consideration: the "Aarhus Convention" – the 1998 Convention on Access to Information, Public Participation in Decision-making and Access to Justice in Environmental Matters and the "Helsinki Industrial Accidents Convention" – the 1992 Convention on the Trans-boundary Effects of Industrial Accidents. The Helsinki Convention explicitly connects with the Espoo Convention stating that where is a hazardous activity subject to Espoo Convention, the final decision shall fulfil the relevant requirements of the Helsinki Convention as well (UNECE 2006).

Table 3 summarizes the ratification of the Helsinki Convention in the SEE region.

Table 3 Review of the ratification of the Helsinki Industrial Accidents Convention in SEE region

Country	Helsinki Convention ratification
Albania	5 January 1994
Bosnia and Herzegovina	–
Bulgaria	12 May 1995
Croatia	20 January 2000
Greece	24 February 1998
Montenegro	19 May 2009
Romania	22 May 2003
Serbia	Under preparation
FYR Macedonia	–

2.2 Bilateral and Multilateral Agreements Under the Espoo Convention

The Espoo Convention encourages Parties to draft more precise bilateral or multilateral agreements to implement their obligation under the Convention by clarifying the more general principles of the Espoo Convention (Espoo Convention, Article 8). By signing additional bilateral or multilateral agreements, countries can overcome difficulties due to the differences between national legislation and EIA practice of the different Parties.

For these agreements, the Espoo Convention even provides recommendation for their contents in its Appendix VI.

The most important elements set out in Appendix VI for the suggested bilateral or multilateral agreement are:

- harmonization of policies and measures for the protection of the environment among countries in order to achieve "the greatest possible similarity in standards and methods related to the implementation of environmental impact assessment";
- establishment of threshold levels and more specified criteria for defining the "significance" of trans-boundary impacts for the geographically specific conditions of the area where the proposed activities and EIA will take place, as well as establishment of critical loads of trans-boundary pollution;
- suggestion to undertake a joint EIA where appropriate.

For the practical implementation, important issues regarding trans-boundary public participation procedure can be also part of such detailed agreements, for example:

- responsibility for organizing public participation;
- time scale;
- financial aspects of public participation;
- translation of materials for the public;
- methods of informing the public and receiving their comments;
- volume and format of EIA materials presented to public;
- methods of informing the public about final decision on a proposed activity, etc. (ECE/MP.EIA/7, Guidance on Public Participation in Environmental Impact Assessment in a Trans-boundary Context, 2006).

Many UNECE countries have taken this approach and concluded such bilateral agreements so far (UNECEb).

3 Implementation of the Espoo Convention in the SEE Region

Countries of the SEE region are relatively small sized, with many similar neighbouring countries they share 13 river basins and four trans-boundary lake basins and have many areas under special biodiversity protection close to the border regions. Because of this geography and large infrastructure developments in the region, a large numbers of projects are likely to cause significant trans-boundary impacts. Additionally, as the Espoo Convention does not only apply to trans-boundary impacts between neighbouring Parties but also to long range trans-boundary impacts, the list of projects falling under the Espoo Convention expands and includes activities with air pollutants or water pollutants, activities potentially affecting migrating species and activities with linkages to climate change (UNECEc).

The practical implementation of the Espoo Convention in the SEE region varies among countries – Bulgaria and Romania are more advanced and experienced in trans-boundary EIA, both as countries of origin and as affected parties, while the other countries have yet to explore compliance with the Espoo Convention. This is especially true with countries that relatively recently ratified the Convention: Bosnia and Herzegovina, Montenegro and Serbia. Turkey is still not Party to the Convention. From the list of trans-boundary EIA cases (UNECEd) in the SEE region, it can be concluded that characteristic projects from the SEE region that are causing or expected to cause significant trans-boundary impacts are as following:

- transit oil and gas pipelines, such as Burgas to Alexandrupolis crude oil pipeline, Gas pipelines between Croatia and Slovenia, Trans-Balkan Oil Pipeline (AMBO), Nabucco pipeline project;
- existing and future nuclear power plants, such as Nuclear power plant at Belene, Nuclear power plant at Chernavoda, Second reactor at Slovenia's Krsko NPP, Nuclear power plant at the Shkoder Lake;
- construction of regional roads and bridges, such as Second Danube Bridge Vidin, Karas River project, the Multipurpose Channel Danube-Sava;
- industrial facilities situated close to national borders or with large cross border impacts, such as Gold-silver and Base Metal Ore Mining in Certej Perimeter, Hunedoara County, Rosia Montana gold mine, Ecological Incinerator Radauti.

3.1 Public Participation Procedure Within Trans-boundary EIA in SEE Region

Public participation is a very important element of the Espoo Convention, which specifies the right of public to be informed and to express its views related to the proposed project. The main challenges for a successful application of the public participation rules in the SEE region are coming from the differences in national EIA legislation, as well as from the poor experience with the EIA tool and the lack of partnerships between government authorities and civil society organizations.

3.1.1 Belene Case

Bulgaria's Belene nuclear power plant project has raised deep concerns in the neighbouring Former Yugoslav Republic of Macedonia. To this respect, the Macedonia-based

CSO Eco-Sense consider that Macedonian citizens should have the opportunity to participate in the project assessment (EIA procedure) and asked the Bulgarian government for the full disclosure of information concerning the project (i.e. the EIA study) and the involvement of citizens in line with the Espoo Convention principles on trans-boundary impact assessment. In this sense, CSO even went to Bulgaria's Supreme Administrative Court, citing violations of both Aarhus and Espoo convention principles and after one year, the Court recognised the duty of Macedonia to disclose the project information, but did not take a decision on citizen's involvement.

When the NGO consulted with the Ministry of Environment and Physical Planning (MoEPP) they found out that the MoEPP discussed the issue with the Espoo Secretariat, and that it was argued at the time that Espoo Convention principles only applied in cases where the distance involved was 100 km or less (the Belene site is located some 275 km from the Macedonian border).

However, in case of the EIA procedure for the Cernavoda NPP, Austria was invited in 2007 to participate in the EIA and ESPOO procedure for the construction of NPP Cernavoda Unit 3 and 4, although not being a border country. The Austrian Institute of Ecology was commissioned to elaborate an Experts Statement to the EIA report. The Report concluded the EIA did not meet the requirements of the EU EIA directive and of the ESPOO Convention as it lacked essential information concerning treatment and interim storage of radioactive waste and its impact on the environment.

3.1.2 "Buk Bijela" Case

This case concerns international waters where the two governments of Montenegro and Republic of Srpska prepared in 2004 a project for the construction of hydro-power plant on the Tara River in Montenegro unilaterally and disregarded the basic principles of international environmental law, such as public participation in the decision-making process and involvement of other relevant stakeholders such as Sava Commission and UNESCO.

The project raised a strong opposition both locally and internationally under the lead of environmental groups from Montenegro which finally stopped the project, albeit temporarily.

3.1.3 Common Problems

Some of the identified problems from the above mentioned cases are:

- the identification of the affected parties, particularly in cases of nuclear power plants; and
- the public participation in the EIA procedure of the affected Party, when the affected Party decides not to participate in the EIA procedure but their citizens wish to take part.

These problems are universal for all countries that are Parties to the Espoo Convention, and from the analysis (INECE) it can be seen that additional understanding and interpretation of public participation rights of the affected Party is needed and practically this question can be solved by countries themselves (e.g. within bilateral and multilateral agreements).

4 The "Small Espoo Treaty"

Countries of the SEE region are participating in sub-regional cooperation and capacity-building activities under the Espoo Convention. Since 2004 till 2008 three workshop meetings were organized in order to strengthen contacts between the Parties, while recently, between the 4th and the 5th Meetings of the Parties (from 2008 onwards), countries met twice firstly in Bulgaria in November 2008 to discuss relationship between EIA and SEA, and then in Montenegro in December 2009 to strengthen the awareness of application of the Espoo Convention. Largely as a result of these meetings, involved countries initiated to draft a multilateral agreement under the Espoo Convention.

4.1 Brief Notes on the Reasons for Signature and Status of the Agreement

Countries of South-Eastern Europe signed a Multilateral Agreement among them for the Implementation of the Convention on Environmental Impact Assessment in a Trans-boundary Context during the Fourth meeting of the Parties to the Convention held in Bucharest, from May 19th to 21st, 2008 (so called "Bucharest Agreement" and "Small Espoo") (UNECEd). The Agreement was initiated by Serbia in 2003 and 2004 to ensure a common EIA procedure concerning major projects between SEE countries, as well as to provide smooth implementation of the Espoo Convention, especially by conducting joint EIA.

Seven countries are signatories: Bulgaria, Croatia, Greece, Macedonia, Montenegro, Romania and Serbia, while Bosnia and Herzegovina and Albania indicated they may join the agreement later. It is important that the non party status to the Convention is not a barrier for accepting the multilateral agreement (e.g. Turkey) (Economic Commission for Europe 2009). Albania also expressed its expectations regarding the ratification of the Bucharest Agreement, during the Twelfth meeting of the WG on EIA of the Espoo Convention in Geneva, on 11–13 May 2009. Turkey, not being a Party to the Convention, indicated that there was not an expectation of accession at present.

The aim of the Agreement is to provide a common EIA procedure concerning major projects/activities between South-East European countries that may have an adverse trans-boundary environmental impact. The Bucharest Agreement shall enter into force on the 30th day after the date of deposit of the third instrument of

Table 4 Status of Bucharest agreement for SEE countries

Country	Signatories of the Bucharest agreement	Ratification
Albania	–	–
Bosnia and Herzegovina	–	–
Bulgaria	20 May 2008	23 January 2009
Croatia	20 May 2008	–
Greece	20 May 2008	–
Montenegro	20 May 2008	16 July 2009
Romania	20 May 2008	–
Serbia	20 May 2008	–
FYR Macedonia	20 May 2008	–

Status: Signatories: 7. Parties: 2

ratification, acceptance, approval or accession. Currently, the Agreement is not yet in force, since there is at least one instrument of ratification needed, in addition to the Bulgarian (approval January 23rd 2009) and Montenegrin (Parliament ratification on July 16th 2009) ones. The ratification of the Bucharest Agreement for Romania is foreseen in 2010 (Sandu 2009) (Table 4).

4.2 Main Provisions

The Bucharest Agreement has eight pages, 24 Articles and one Annex (UNECEd). The Parties to the Agreement committed themselves to take all the necessary legal, administrative and other measures to implement the provisions of Espoo Convention (Article 3) as well as to implement the provisions of the Agreement in regard with the proposed activities listed in Appendix I of the Convention that are likely to cause significant adverse trans-boundary impact (Article 4). The question on whether the activity will undergo a trans-boundary EIA procedure or not is to be answered by the Parties themselves who are obliged to firstly develop and adopt in future (likely in the form of an Annex to the Agreement) a set of criteria for the identification of significant adverse trans-boundary impact, based on the general criteria set forth in Appendix III of the Convention (Article 5). Appendix III proposes to consider the size of the activities (which are large or not for the type of the activity), the location (which are located in or close to an area of special environmental sensitivity or importance) and the effects (particularly complex and potentially adverse effects).

For the criteria on significance, the Parties will seemingly use additional criteria and threshold levels set out in the IPPC and Seveso II Directives. Once the criteria are set, the Parties will further specify and interpret other important issues and terms of the Convention, in form of guidelines on the implementation of the Agreement. The guidelines will be based on the following elements, amongst others: screening, notification, confirmation of participation, transmittal of information, preparation of environmental impact assessment documentation and its distribution, public participation, consultations between Parties, decision and transmittal of final decision, post-project analysis and translation (Article 5.2).

When there is a special case in applying the Convention, e.g. joint cross-border projects ("joint proposed activity" under the jurisdiction of two or more states), the Agreement envisage in its Article 6 that the Parties concerned shall conduct EIA public consultations and communication according to arrangements determined by one or more joint working groups.

Also, if Parties so agree, a number of provisions of the Agreement (e.g. Articles 7 to 11) might not be implemented as well (Article 6.2).

These provisions are related to:

- notifying the affected Party without undue delay about the proposed activity falling under the Appendix I of the Convention by the Party of origin (Article 7.1.);
- notifying the affected Party at scoping stage (if exists), or earlier, by the Party of origin (Article 7.2);
- response (to the Party of origin's notification) from the affected Party in 30 days timeframe indicating willingness to participate in trans-boundary EIA (Article 7.3);
- information content of the notification, as prescribed by the Annex of this Agreement (Article 7.4);
- procedure in case of negative or late response (to the Party of origin's notification) of the affected Party, which states articles 9 to 14 will not be applied in such case (Article 7.5);
- translation of the notification in English by the Party of origin (Article 8.1);
- duty of the affected Party to translate information related to the potentially affected environment and comments received by public and authorities, if necessary (Article 8.2);
- right of the affected Party to request subsequent communication and EIA report in English (Article 8.3);
- determination of documentation to be translated by the project proponent into the official language of the affected Party (Article 8.4);
- possibility of establishing one or more joint working groups for subsequent communication and the exchange of information between the concerned Parties (Article 9);
- joint decision of concerned Parties on the details of the arrangement for the distribution of EIA documentation to the affected Party such as number of copies, location and timing (Article 10.1.a), as well as the way of submission of public comments of the affected Party (Article 10.1.b);
- determination of minimum information content of the EIA documentation, as described in Appendix II of the Convention (Article 11).

Within a normal procedure, foreseen by the Agreement, the involved Parties need to specify in the final EIA decision how the public comments of the affected Party will be taken into account (Article 12), and whether there is a judicial procedure of the Party of origin to challenge the final decision (Article 13). Post-project analysis or monitoring according to national legislation also might be agreed among the competent authorities of the involved Parties (Article 14). Finally, it is on the Parties to undertake "without undue delay" consultations and mutually agree on whether the proposed activity is an activity listed in the Appendix I of the Espoo Convention (e.g. whether there is likelihood of significant adverse trans-boundary impact – Article 15).

If there is a written request for a meeting of any Party supported by at least one other Party, the meeting will be arranged within 90 days (Article 16). Whit regard to the amendments to the Agreement, Article 18 states that proposed amendments might come from any Party in writing form. The agreement also envisages the possibility for withdrawing from it, by giving written notification to the Depositary (the Government of Romania), at any time after 2 years from the enforcement (Article 23).

4.3 Some Discussions Related to the Bucharest Agreement

4.3.1 Public Participation Rights

The Bucharest Agreement is the first multilateral agreement signed under the Espoo Convention, in addition to the number of bilateral agreements signed so far. It may contribute also to the on-going EU accession process for SEE countries and to align national EIA procedure with the EU EIA Directive (which incorporates Espoo Convention's and other UNECE Convention's provisions). (Bogdanovic 2008).

As countries on the Balkan are geographically located today in such a way that large water resources and pathways overlap borders – there are now 13 internationally shared river basins and four trans-boundary lake basins – there is a rationale for the agreement that facilitates the implementation of the EIA in a trans-boundary context. Since the Espoo Convention itself has a quite vague language, there is a need for interpretation of various terms like "significant" impact, "major change to an activity", "reasonably obtainable information", "reasonable alternatives" as well as to address practical arrangements for public participation and the procedural steps. In all cases, purpose of a bilateral agreement is to enable the effective application of the Convention, especially in case of joint projects, where countries involved are considered both as the Party of origin and the affected Party.

For a "joint proposed activity", the Bucharest Agreement however gives the right to all concerned Parties to skip relevant procedures and practical arrangements for disclosure of information, public consultation and communication set under this Agreement and conduct EIA public consultations and communication according to special (unknown) arrangements.

These arrangements are to be the product of one or more joint working groups. If case by case consultations would be provided for joint projects activities that will take place under the jurisdiction of more than one Party (e.g. nuclear power plants, hydropower plants, cross border pipeline etc.), this would actually reduce standards for public participation making it more difficult, inaccessible and non transparent than before. Thus, the whole spirit of the Espoo Convention is being omitted, as the Convention promotes bilateral or multilateral agreements with even more stringent measures than those of the Convention (Article 2.9), and as the provisions of the Espoo Convention shall not prejudice any obligations of the Parties under international law (Article 2.10), in this case the Aarhus Convention.

In the SEE region there are already some problems identified with the public consultation processes within trans-boundary EIA, and detailed arrangements for public participation (responsibility for organizing public participation, time scale, financial aspects of public participation, translation of materials for the public, methods of informing the public and receiving their comments, volume and format of EIA materials presented to public, methods of informing the public about final decision on a proposed activity) should be determined as clear as possible in multilateral agreements aimed at enabling a full compliance with the Espoo Convention. The recommendation on NGOs, stated in Decision (II/8) on strengthening sub-regional cooperation, adopted during the Second Meeting of the Parties in 2001 to strengthen the role of non governmental organisations in implementation of the Convention was not used in drafting the Bucharest Agreement neither.

4.3.2 Trans-boundary Aspects of Biodiversity Protection in SEE Region

The SEE region is still rich in biodiversity and these countries, now Candidates and Potential Candidates, once in the EU will increase the EU biodiversity. Many areas under special protection are located in, or close to, border regions.

However, because there are on going processes of degradation of natural assets, there is much to be done in order to avoid irreversible effects. During the 9th Conference of the Parties to the Convention on Biological Diversity (IUCN 2009), six SEE governments (Albania, Bosnia and Herzegovina, Croatia, Montenegro, Serbia, and Slovenia) announced their intent to declare 13 new protected areas and enlarge nine existing ones. The Balkan countries have yet to declare and protect legally areas and sites according to the EC Directives 79/409 and 92/43 (SPAs and Natura 2000 sites respectively).

Relevant international and EU legal instruments are more stringent than those of the Espoo Convention whit respect to biodiversity protection. The interest of South Eastern European countries to protect rich biodiversity has to be considered as an integral part of national EIA legislation and bilateral EIA agreements. When developing general criteria to determine which activities can lead to significant trans-boundary adverse impacts, as well as specific scoping criteria, countries of SEE region need to take into account the relevant guidance developed under relevant treaties such as Ramsar Convention guidelines on the wise use of wetlands, Biodiversity Convention COP Decisions, guidelines on trans-boundary EIA under the Convention on the Protection of the Black Sea, Red List of Endangered Species and existing bilateral agreements related to the biodiversity conservation and management of shared nature protection areas (like the Memorandum of Understanding in the field of environment protection and sustainable development signed between

Albania and Montenegro and the MoU on protection of the Stara Planina region between Macedonia and Bulgaria).

5 Conclusion

Countries of the SEE region should take advantage of sub-regional cooperation under the Espoo Convention to develop its civil society and democracy by setting up common and clear mechanisms for public participation in decision making on trans-boundary projects. Also, biodiversity, conservation and management of shared nature protection areas in the SEE region should be placed and emphasised in the trans-boundary EIA regime specified in the Bucharest Agreement.

Taking into account the current and potential loopholes of the Bucharest Agreement, an action is needed to reconsider the potential diversion of the Bucharest Agreement from the Espoo Convention and other UNECE Conventions. Since "the enforcement mechanisms in the Espoo Convention are not particularly strong" (Cassar and Bruch 2004), and NGOs have already urged for a stronger role of the Implementation Committee of the Espoo Convention and for the commitment from the Parties to allow communications from the public to the Implementation Committee, the loopholes of the Bucharest Agreement should be addressed not only by the NGO community but primarily by the Parties themselves.

References

Andrusevych A (2009) Statement of the European ECO Forum to the 3rd Meeting of the Parties of the Espoo Convention, http://www.unece.org/env/eia/documents/cavtat/Andriy%20Andrusevych.pdf

Bogdanovic S (2008) Legislative Aspects of EIA - Current & Development Trends in the Danube Basin, 37th IAD Conference, Chisinau, Moldova, 29.10-01.11, 2008

Cassar AZ, Bruch CE (2004) Trans-boundary Environmental Impact Assessment in International Watercourse Management, 2/10/2004, www1.law.nyu.edu/journals/envtllaw/issues/vol 12/1/12n1a7.pdf

Construction of NPP Cernavoda Unit 3& 4, Experts Statement, GZ BMLFUW-UW.1.1.4/ 0029-V/6/2007, www.ecology.at/files/ berichte/E22.572-1.pdf

Economic Commission for Europe (2009) Subregional cooperation and capacity-building, Summary of workshop findings ECE/MP.EIA/WG.1/2009/3, March 2009, http://www.unece.org/env/documents/2009/eia/wg.1/ece.mp

Economic Commission For Europe, http://www.unece.org/env/pp/ratification.htm

Energy Community Secretariat (2008) Report on the implementation of the *aquis* under Title II of the Treaty establishing the Energy Community, November 2008, http://www.energy-community.org/pls/portal/docs/220177.PDF

European Commission (2009) Respective countries' EC 2009 Progress Reports http://ec.europa.eu/enlargement/press_corner/key-documents/reports_oct_2009_en.htm

GRID-Arendal, http://www.grida.no/publications/vg/balkan/page/1370.aspx

INECE. Furlop Sandor: Problems of Trans-boundary Environmental Impact Assessment, Fifth International Conference on Environmental Compliance and Enforcement, www.inece.org/5thvol1/fulop.pdf

IUCN. Article: Historic agreement force for biodiversity in south-eastern Europe, 29 May 2008, http://iucn.org/about/work/programmes/business/bpp_news/?1061/Historic-agreement-for-biodiversity-in-South-Eastern-Europe

REC (2003) Comparative Analysis of International EIA Requirements Relevant to SEE Countries, 2003 http://archive.rec.org/REC/Programs/EnvironmentalAssessment/pdf/EIA-Internationall Req.pdf

Sandu C (2009) Subregional Workshop on raising awareness of application of the Espoo Convention, Podgorica, December 2009, http://www.unece.org/env/eia/documents/ActivityReports/PodgoricaDec09/Presentation_by_Romania.pdf

The Magazine of the Regional Environmental Center Article: Belene project 'fallout' reaches FYR Macedonia, http://www.greenhorizon-online.com/index.php/Former-Yugoslav-Republic-of-Macedonia/belene-project-fallout-reaches-fyr-macedonia.html

UNECEa, http://www.unece.org/env/eia/bucharest_agreement. html

UNECEb, Text of the Bucharest Agreement. http://www.unece.org/env/eia/documents/bucharest/SEE_multilateral_agreement_final.pdf

UNECEc, Convention on the Trans-boundary Effects of Industrial Accidents, Art. 4, para. 4. http://www.unece.org/env/documents/2006/teia/Convention%20E%20no%20annex%20I.pdf

UNECEd, http://www.unece.org/env/eia/resources/agreements.html

UNECEe, Guidance on the Practical Application of the Espoo Convention. http://www.unece.org/env/eia/guidance/practical.html

UNECEf, Reports from Subregional cooperation workshops for South-Eastern Europe under the Espoo Convention, 17-19 November 2008, Koprivshtitsa (Bulgaria) and 15-16 December 2009. Podgorica, Montenegro, http://www.unece.org/env/eia/documents/ActivityReports/PodgoricaDec09/Workshop_report.pdf and http://www.unece.orgenv/eia/documents/ActivityReports/KoprivshtitsaNov08/Bulgaria08_workshop_report.pdf

United Nations Treaties Website, http://treaties.un.org

Promoting Environmental Protection Through the Management of Shared Natural Resources Between Albania and Montenegro: The Shkodra Lake Watershed

Djana Bejko

Abstract The world's watersheds are under an increasing pressure. Environmental security is an important topic which concerns nowadays the quality life of the communities all over the world. It becomes more thematic with respect to natural resources protection and more over freshwater ecosystems shared between bordered countries.

This article evaluates the existing freshwater concerns of Shkodra/Skadar Lake ecosystem as well as the environmental security concern of the local population living in both sides. In fact, there are untreated sewage waters, untreated industrial pollutions and pesticides that are provoking low security of the environment and the water quality of the Lake watershed. This paper offers a model of cooperation in the cross-border context, recommendations for future treating measures toward water quality which can provide environmental security and improving the bio-life quality for Shkodra/Skadar lake ecosystem. Through this paper we argue that ecosystem-oriented principles are essential to shape effective freshwater using priorities as they involve along a continuum from dialogue and partnership, to sharing of information, to more defined frameworks of cooperation at a trans-boundary level, to binding legal norms.

The process we experienced provides also the instruments and mechanisms established at the trans-boundary level and the implementation process mainly focused on the best practices to mitigate environmental problems, institutional strengthening and capacity building running throughout the programme "Promoting Environmental Protection toward Management of Shared Natural Resources between Albania and Montenegro". It counts also on the adoption of environmental legislation with respect to international regulation implementation and related strategies (AKM/NEA 1999).

Keywords Shkodra/Skadar Lake • Shared freshwater resources • Environmental security • Sustainable use • Trans-boundary cooperation

D. Bejko (✉)
Biology Department, The Regional Environmental Center Albania and University "Luigj Gurakuqi", Shkoder, L. Qemal Stafa, Rr. Vasil Shanto, Nr 21, Shkoder, Albania
e-mail: Dbejko@rec.org

1 Introduction: The Lake Shkodra Watershed

Lake Shkodra/Skadar is a trans-boundary water ecosystem between Albania and Montenegro. It is the largest lake on the Balkan Peninsula in terms of water surface (Fig. 1). The drainage area of the lake is about 5,500 km^2. The lake area varies between 353 km^2 in dry periods and 500 km^2 in wet periods. The lake volume varies between 1.7 km^3 in dry periods to 4.0 km^3 during wet periods. The distance between the mouth of the Crnojevica River (northwestern lake edge) and the lake's outlet (Buna-Bojana River) is 44 km (maximum length); its greatest width is 13 km. The most important tributaries of Lake Shkoder enter the lake from the north: Moraca, Crnojevica, Orhovstica, Karatuna, Baragurska River in Montenegro, and Rrjolli and Vraka Rivers in Albania. The area targeted is the home to some 350,000 people, living in five municipalities in Albania and Montenegro (Ziu and Bejko 2004).

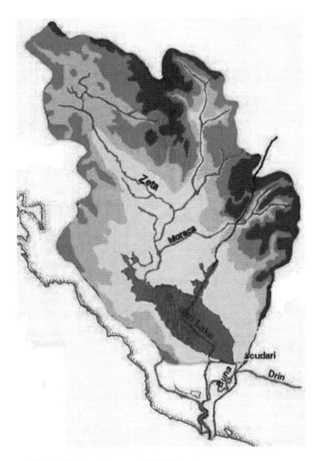

Fig. 1 The basin of lake Shkodra (Source: HMI Tirana)

This freshwater ecosystem is well-known for its diverse set of different habitats with unique biodiversity of flora and fauna and has been recognized as international important ecosystem and waterfowl habitat by proclaiming it a RAMSAR site in both Albania and Montenegro sides (Anonymous 2006). The number of waterflow is determined by winter counts since 1990. Between 1990 and 1999 numbers varied between 150,000 and 250,000, but since 1999 numbers have dropped strongly to 35,000 in January 2005. Shkodra/Skadar Lake has a high variety of fish fauna, the result of a good communication with the sea, and of an extensive network of rivers and streams. Its ichtyofauna includes highland coldwater fish species, warm freshwater fish species and several marine species; in total about 60 fish species belonging to 17 families (Dhora Dh 2005). The relatively high number of endemic species makes the lake significant on regional level. About ten species are commercially exploited (e.g. carp, bleak and eel). Two fish families are especially important: cyprinids (most abundant in species) and salmonids (which are much rarer). In Shkodra/Skadar Lake about 64 genera with 310 species of phytoplanton are met mainly composed by *Diatomeae*. Among the diatoms of the lake, the most common species are *Cyclotella ocellata* and *Aulacoseira ambigua*. Aquatic macrophytes are also meet in the Lake and there are 164 species belonging to 66 genera and 43 families (Kashta et al. 1997). Reed beds and other macrophytes have purification capacities through nutrient retention and transformation (nitrogen, phosphor), and binding of pollutants.

This large variety of habitats supports a rich assemblage of species: 726 vascular plants, with more than 30 rare species, about 60 fish species (15 of them endemic to the water system), 15 amphibians, 30 reptiles, 271 bird species (90% of them migratory species of international conservation concern), and 57 mammals (Dhora Dh and Sokoli 2000).

The Shkodra/Skadar Lake ecosystem is an important source for the local population in both Albania and Montenegro countries. Due to the demographic movements and human activity pressure increasing, the exploitation of natural resources has been intensified after 1990 affecting the environment conditions in general and the Shkodra/Skadar lake ecosystem in particular in terms of Lake basin degradation and water quality concerning (Bejko and Bushati 2008).

The degradation of the Trans-boundary Lake Shkodra basin remains a concern due to unsustainable use of its resources and results from ineffective management schemes and current practices using in managing in trans-boundary level. Among main risks are the following ones: sewage and industrial pollutions, degradation and loss of biodiversity as well as depletion of fish quantity (Bejko 2006).

With the overall goal to help solving environmental problems and strengthening cooperation between the countries of South Eastern Europe in the concrete context of nature protection and biodiversity conservation as a common threaten of the Lake watershed, the pilot cross-border site Lake Shkodra\Skadar was involved in the Regional Environmental Center for Central and Eastern Europe (REC) initiative, where local communities were facing particular development challenges.

2 Working Methodology

The working methodology was conducted through desk research and field work, focusing specifically on water protection legislation at national and international levels, plans, studies and strategies existing at national level in both Albania and Montenegro as the main target area of this study. The study part of the work consisted in the review of documents available on site, gaps and missing studies. A data base was established within the main purpose of addressing environmental concerns in the Lake Shkodra\Skadar Watershed.

Re-establishing trans-boundary dialogue and mutual trust after military conflicts and enabling local actors to manage shared natural resources in a sustainable way were the main initiative tasks. Dialogue had to be launched to involve partners and countries whose cooperation is a key prerequisite for addressing resource management issues in a trans-boundary context. The groundwork needed to be laid so that such a process involving many smaller-scale processes could become sustainable and contribute to achieving the longer-term goal of the project: the sustainable management of shared natural resources (REC 2007). In addition to dialogue between and among countries and communities on the two sides of the border, the integration of local communities into national processes has been important. Efforts were made to improve cooperation between local and national level governments, and to include cross-border sites into national strategic documents and processes related to biodiversity protection and the environment. The nature of this process implementation is unique in the region, and therefore there are no neighbouring countries from which to learn. As a consequence, the project's methodology has been kept as much open and flexible as possible and tailored to the specific needs and circumstances of each site.

2.1 Methodology Elements Used in the Process

Water issues are often simultaneously trans-boundary and local by nature, resulting in unsymmetrical rights and responsibilities that form the basic dilemma in protection and management in terms of water quality and sustainable use of resources. The relationships between upstream activities and downstream impacts are very complex, further complicating the Albania-Montenegro inter-governmental negotiation processes of water governance. In this context, different elements on working were elaborated and implemented as the following:

- Relying on an international cooperation framework to ensure national ownership.
- Treating the initiative as an open-ended process – building each step on the one before.
- Fostering local participation by engaging as many relevant stakeholders as possible from the very beginning.
- Identifying priorities locally – working with proposals provided by local stakeholders based on the information and experience they collected themselves.

- Establishing permanent trans-boundary bodies to engage stakeholders in the joint planning of activities and maintaining cross-border dialogue.
- Rallying communities around a joint vision for the Lake Shkodra/Skadar region.
- Empowering local actors to become leaders of trans-boundary processes.
- Allowing stakeholders to take action and learn from their own results by providing direct support to pilot local initiatives.
- Promoting a positive regional image through trans-boundary promotional activities and by bringing communities together.
- Ensuring transparency through a systematic approach to communication.

2.1.1 Ensuring National Ownership

Without the political will and the ownership of national governments, it would have been impossible for the project to address issues of trans-boundary concern. Even if the problems addressed were mostly of local character, they often interfered with national level processes – new strategies, designation of protected areas – and required national level participation. The initiative was running under the REReP Programme for South Eastern Europe, a framework built on the strong political will of countries under the Stability Pact for South Eastern Europe and to integrate environmental concerns in the process of their reconstruction and development. In this context, the initiative had the trust and support of all of the national governments involved. They expressed their commitment by nominating ministerial focal points to act as key contacts for the project and by providing direct contributions in political, financial and professional terms.

The increasing commitment of local actors also contributed to mobilization of national ownership. Results achieved in the three cross-border sites often led to national level commitments concerning the future development of the sites, and locally identified priorities were included into national strategic planning documents.

2.1.2 The Initiative as an Open-Ended Process

The running process in the site of Lake Shkodra/Skadar was rather diverse and experienced the difficult local context of a cross-border site. It addresses simultaneously various levels of decision making, and, in addition to some very concrete end-results, it seeks to maintain a sustainable process. The sustainable management of natural resources and preserving natural and cultural heritage while delivering benefits to local communities are in the heart of the consistent drivers. At the same time, the initiative runs in parallel with concrete local development processes. Besides supporting these processes, it aims at building on the new opportunities that arise and seek to help local actors to utilize them. These efforts require regular planning, which involves local stakeholders, as well as open possibilities to include new activities in response to new opportunities or specific local interests.

2.1.3 Fostering Local Participation

Local participation and ownership of local stakeholders have been both the goal and a precondition for any success. The role of process driving is to neutrally facilitate in relation to local stakeholders engaging in processes of cooperation. To ensure these results, trans-boundary dialogue was initiated with the widest possible scope.

Local kick-off events – "get everyone involved meetings" – and regional planning meetings involved practically anybody interested, engaging actors on both the national and local level from Albania and Montenegro. In Shkodra\Skadar Site-level, discussion platforms resulted in the nomination of representatives for cross-border regional planning working groups, which brought together over 110 stakeholders from both Albania and Montenegro countries including representatives of government agencies, local authorities, research and business communities, schools, and civil society. Local participation was initiated and guided by several fundamental principles:

- Partnership and participation frees people from dependence on old, authoritative and centralized governance models.
- Partnerships and cooperation, features of stakeholder participation, are characteristics of an emerging civil society, and can substantially strengthen democratic processes.
- Broad-scale stakeholder participation creates conditions for community ownership, openness to capacity-building activities and empowerment.

To maintain participation, each step of the process was decided based on the joint analysis of the achievements and identification of needs. Strong local involvement from the very start enabled stakeholders from both Albania and Montenegro to come up with a strong vision and establish priorities they believed in.

2.1.4 Identifying Priorities Locally

Once local participation was initiated, further efforts were needed to keep it alive. Several methodologies have been used for this, one of them being the joint definition of priorities for local development and trans-boundary cooperation. The lack of knowledge about biodiversity and natural resources, the unavailability of sufficient and up-to-date information and the lack of unified approaches of collecting biodiversity and environmental data were perceived as key obstacles for the development of joint management. As a response, teams of local and national experts from neighbouring countries worked together on collecting and analyzing available information. Joint studies were elaborated, which served as a direct input for continued trans-boundary dialogue and the defining of local priorities. Common methodologies for assessing shared resources, the practicing of local experts in dealing with local environmental issues and the availability of new information for the benefit of management authorities were important by-outputs of this process.

Still, the overall purpose was not to deliver maximum quality scientific studies, which perhaps could have been well performed by international consultant teams. Instead, the goals were to get relevant local players involved and trained on the process, and to ensure local inputs to trans-boundary debates.

2.1.5 Trans-Boundary Dialogue and Joint Actions Planning

Permanent trans-boundary forums represented another key methodology for maintaining local participation and transparency in the project. Forums composed of key local stakeholders from both sides of the border – NGOs, local authorities, schools, tourism boards, water management and nature conservation authorities, environmental inspectorates – directly contributed to defining and planning activities (Fig. 2). To ensure their involvement, regular forum meetings have been organized four to five times a year.

Additionally, the process coordinators, as the main facilitators on the local level, Shkoder and Podgorica, maintained close cooperation with forum members. As an informal body, forums were also the first seeds of institutionalized trans-boundary dialogue. They developed at a different pace in different sites – still following a common concept – and in Albania and Montenegro they became legally registered bodies to promote trans-boundary cooperation. Forums role and scope of work significantly expanded from initially being a process planning and a consultation mechanism to becoming real actors in the planning and implementation of regional and cross-border development policy. From being representative bodies of local actors and communities, forums evolved into groups that developed strategic approaches for their whole trans-boundary region.

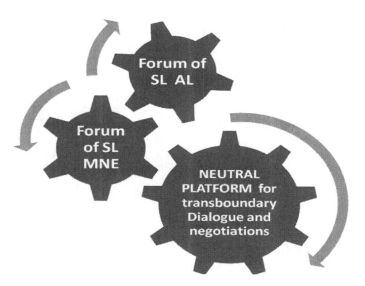

Fig. 2 FORUM of Shkodra ake – mechanism for dialogue and partnership

2.1.6 Joint Trans-Boundary Vision Developing from Cross-Border Communities

A key characteristic of the project is that it is community-based. Even though government authorities contributed authority and influence, the project itself was driven by local communities sharing a sense of ownership of common natural resources. This attitude proved to be strong enough to overcome separation by borders and alleviate cross-border tensions. Regardless of the country they were from, people became members of new cross-border communities united by the awareness that the natural entity that they rely on is common, and it is their joint responsibility to preserve and manage it in a sustainable manner (Ervin 2003).

These attitudes were clearly expressed when members of local communities engaged in defining their common trans-boundary vision concerning the sites. The planning of future activities was directly based on these vision statements. After several years of implementation, towards the end of the second phase, site level processes significantly matured: priorities and main directions for future action have been defined, and the first results encouraged all involved in the project. Site coordinators fully integrated into local networks of key players, and trans-boundary forums became well established and accepted. At this stage vision statements of the three trans-boundary communities were revised reflecting the internal evolution of the project. Site coordinators and the project team worked together with local stakeholders to re-set priorities and define objectives for activities for the following 5 years – 2003–2008 – and beyond. Although these were to serve the development of project action plans for the third phase, they also provided an opportunity to assess results and engage local actors in further defining their roles for the future. As an end-result of this process, a diagnosis of the current situation was carried out and concrete action plans were developed for all the three sites.

2.1.7 Empowering Local Actors

Empowering of local actors, means raising their awareness of issues related to the management of natural resources, building their capacities for taking action and increasing the understanding of specific development opportunities they have. Members of local communities often did not appreciate the natural abundance of their region, nor the development potential this provides. Their abilities and motivations to participate in the project as partners were therefore limited at the beginning. The new types of cross-border relationships envisaged required new skills and capacities that had to be developed during the initial period of implementation. The need for intervention was clear when sites were selected based on their initial appraisal, but this was understood by a wider range of actors only when issues like nature conservation, the sensitivity of natural resources or the need to preserve them had been raised. To improve capacities of local stakeholders, several strategies were pursued, including:

- Disseminating knowledge and increasing the understanding of natural and cultural values through topic-oriented training for local stakeholder groups.
- Developing site-specific solutions to address nature conservation problems together with affected stakeholders and with the application of their traditional knowledge.
- Providing information on alternative approaches to the use of natural resources: organic agriculture, sustainable tourism, and linking nature conservation with agriculture and rural tourism.
- Assisting local players in developing their initiatives into concrete projects and in raising additional funds for their implementation.
- Establishing links between organizations dealing with conservation and management to facilitate the exchange of experience and the transfer of best practices across borders and among project sites.
- Strengthening the capacities of local and national decision makers to develop and implement action plans for the management of protected areas and natural resources and.
- Training of trainers, with focus on building and retaining knowledge and expertise on the local level that could be further disseminated to entire communities.

2.1.8 Direct Support to Pilot Local Initiatives

As a locally rooted process that seeks to provide benefits locally, initiatives by stakeholders were treated as expressions of local needs that require attention and support. The methodology was based mainly on concrete local processes, which – with some additional support from the project – reached their completion. In this way local actors could not only deliver something beneficial to their communities, but they could also learn and reap the benefits of taking action. Support to civil society has been a special component involving several rounds of small grants distribution. Projects of NGOs and other non-profit organizations addressing various topics have been supported, with the aim to strengthen their – and engage them in – nature conservation, environmental education and public participation. The grants scheme shifted from an open granting procedure to targeted grants, where a relatively larger amount of funds was given for well articulated projects developed by local stakeholders in cooperative way of Albania-Montenegro NGOs applying together.

2.1.9 Bringing Communities Together – "People-to-People Approach"

A set of activities has been developed to build relationships not only among institutions representing local communities, but also among community members themselves. Large-scale promotional events have been dedicated to this purpose. They brought local people together, promoted the general values behind the project and helped develop positive regional images. Building new relationships among societies divided by country borders but sharing natural surroundings was a central feature of these initiatives. Many of them have become new traditions; local and, in some cases national, governments have lent support to perpetuating them as annual events mentioning here "day of shkodra/skadar lake".

2.1.10 Ensuring Transparency

To ensure the transparency, in addition of regular open discussions with stakeholder, project newsletters, bulletins, websites, promotional materials and other means were employed to reach stakeholders involved in the process and the wider public in both Albania and Montenegro countries. The regular, results of the project were always distributed using mailing lists created and updated continuously during the process implementation. Cooperation with local media proved very important and its presence was ensured at every event. In return local newspapers, TV and radio stations communicated important information about the project. The main methods used to ensure transparency include:

- Exchange of information at cross-border level: the process implementation was weekly reported among project units in local, national, cross-border levels. The exchange of information concerning developments and deliverable was also maintained through regular team meetings.
- Publicity of project actions and achievements: a public communication strategy was developed in cross-border level. Awareness programmes tailored to local needs and mailing lists with over 100 entries have been maintained, including contacts of key international organizations and authorities in bilateral level. Fact sheets, newspaper articles, research results, announcements and other information were regularly disseminated. TV spots, presentations, campaigns and promotion materials were used to keep players informed.
- Communication with national institutions and other international players: national governments in both Albania-Montenegro countries were regularly informed through communication with focal points established within the respective Ministries of Environment. Information about the progress was also disseminated to other international organizations and donors, allowing them to join the process and adding value through their own initiatives. The information disseminated and provided at international level stimulated international initiatives in the site like the European Green Belt of IUCN, the World Conservation Union; and the Dinar Arc Initiative of UNESCO, WWF, UNDP, IUCN and the Council of Europe (REC & SDC 2001).
- Regular reporting to the REReP Task Force: the project and its achievements have been presented to governments and international donors at task force meetings of the Regional Environmental Reconstruction Programme (REReP).
- Proactive communication and openness of processes at the site level: the spirit of openness has been created through the very first site level actions and has been kept throughout the whole process. Working groups, meetings of stakeholders and trans-boundary forums all followed the principle of transparency.

The transparent character of this process had an important secondary effect. The high involvement of the media and actions publicity led to an increased feeling of responsibility of stakeholders participating in the process. The awareness of the fact that the entire community was following the work inspired participants to do their best.

3 Key Process Outcomes

The whole process had tangible results and achievements which brought outcomes at trans-boundary level like the following:

3.1 Communication Between and Among Institutions and Countries

Working relations between central governments and local authorities in both Albania and Montenegro have been strengthened, and their joint decision making has been encouraged by engaging them in joint (trans-boundary) planning activities. Such forms of cooperation were nearly non-existent before.

For the first time after 90th, local stakeholders from Albania and Montenegro countries were brought together to define common objectives and activities for the future of their communities. Regional planning groups composed of representatives of national governments, authorities of protected areas; local authorities, NGOs and the private sector have been established. After years of isolation (especially Albania), operational contacts were established between important national level players, including the ministries of environment in Albania and Montenegro. These working relations remained alive throughout the project, expanded, and got strengthened by official declarations, MoUs, as well as Statements of Cooperation.

Substantial national and international attention has been mobilized and the cross-border site of Shkodra/Skadar Lake that was mostly neglected in past years was included in national strategic documents on the environment.

3.2 Operational Cross-Border Cooperation Mechanisms

The Trans-boundary forum regularly brought together key stakeholders and experts, allowing them to follow developments within the project and the region, became an important institutional basis for future trans-boundary cooperation. As extensions of forums, topic-oriented trans-boundary working groups were established, targeting topics like nature conservation, biodiversity protection, sustainable monitoring, eco- tourism, agriculture and/or education.

3.3 The Introduction of Participatory Processes

In overall the methodology of the project, its openness and bottom-up character introduced new participatory methods of work to local stakeholders and institutions. This process offered many opportunities for both Albania-Montenegro stakeholders to learn how cooperation with others working in the same local context can help them to deliver better results within their own field of work. Trainings and capacity building actions on public participation issues and participatory methods, like Strategic Environmental Assessment and Environmental Impact Assessment strengthened these lessons with more theoretical background and knowledge.

3.4 Trans-Boundary Cooperation Agreements

The work of trans-boundary forums receives the important support of official trans-boundary cooperation agreements where local actors and national governments clearly engage to maintain and strengthen these newly established forms of work.

In order to strengthen cooperation concerning the management and protection of the Shkodra/Skadar Lake national Ministries of Environment in Albania and the Republic of Montenegro signed a memorandum of understanding (MoU), which has been followed by launching their joint initiative to celebrate the "Day of the Lake Shkodra" every year through cross-border events.

The Municipalities of Shkodra, Albania and Cetinje, Montenegro went through the process of developing their Local Environmental Action Plans (LEAPs), and as one of the outcomes, they issued their official statement that lays down the main principles for their cooperation.

3.5 Designation of New Protected Areas

The project made substantial contributions to the designation of new protected areas, which would ensure that nature conservation efforts on one side of the border are complemented with similar ones on the other. Through this, important elements of a harmonised approach to the protection of shared ecosystems are getting prepared.

Following the decision of the National Assembly of Albanian, shores of the Shkodra Lake and adjacent wetland areas (23,027 ha) have been assigned the status of "Managed Nature Reserve" – IVth IUCN category and were included in the national system of protected areas. The same area received international recognition by receiving the status of wetland of international importance under the Ramsar Convention (Anonymous 2006).

3.6 Contribution of Joint Studies

The process started up in empty ground in terms of watershed protection and conservation researches and studies. The main studies developed under this project have been joint initiatives and it was used the same methodology and structure for the studies carried out. Such studies were based on joint Albania-Montenegro working teams addressing the management of shared natural resources. Joint bibliographies of relevant literature, biodiversity surveys and databases, analysis on different aspects of water management, use of forests, environmental awareness of local communities, roles and responsibilities of key stakeholders help local planning and decision making in a trans-border context.

3.7 Developing Common Strategies

Defining common priorities, planning joint actions and developing a vision and objectives has been an overarching element of the whole project. Local stakeholders from Albania-Montenegro were engaged in such processes from the very beginning. Through the process they received support in elaborating their own visions and in defining directions in more concrete areas of work. Apart from participating at various planning meetings and defining common priorities based on the information collected by them, stakeholders engaged in developing strategic concepts on areas like sustainable tourism and agriculture or environmental protection.

Halfway through the project a synthesis of existing knowledge has been prepared in a trans-boundary context, along with a diagnosis of main problems and opportunities and 8 year strategic plans. Direct support to local initiatives of local stakeholders provided opportunities to test how activities identified together can work in real life situations.

3.8 Knowledge and Experience Sharing

Key local stakeholders benefited through the project from a wide range of targeted capacity building actions – trainings, workshops and exchanges – on topics like objective oriented planning, Local Environmental Action Plans, Environmental Impact Assessment, Aarhus Convention, proposal writing and project cycle management, biological monitoring, habitat typology, education programmes and curriculum development, sustainable tourism and agriculture and sustainable hunting.

Stakeholders participated in several exchange programmes, and learned from the experiences of communities on the other side of the border and similar areas in countries of the region and beyond. Study visits have been organised to areas like the Ohrid Lake, between Macedonia and Albania, Peipsi Lake, between Estonia and Russia, the protected Lonjsko Polje wetland in Coratia, Constance Lake among Switzerland, Germany and Austria, Abruzzo National Park in Italy or the Delta of the river Po as well as Danube River Delta in Romania.

3.9 Management of Shared Natural Resources

New methods for addressing issues related to the management of natural resources on the level of local communities have been introduced. Forum meetings, topic oriented working groups and workshops on concrete resource management aspects provided opportunities for stakeholders to elaborate their joint approach towards issues like fishing or hunting in protected areas, adopting environmentally friendly farming methods, or developing sustainable tourism and so on.

In the framework of direct technical assistance management authorities, local authorities, agricultural bodies and schools received basic equipment for office operations and monitoring, education facilities and other tools necessary for performing their work.

Improved capacities of management authorities and benefits of a multi-stakeholder approach come together in concrete local initiatives where, for example, processes aiming at the designation of new protected areas as well as, the development of common tourism products or safeguarding agro-biodiversity are implemented in a joint effort.

3.10 The Way Towards Stability and European Integration

The project – due to its scope and nature – aligned with the many other initiatives intending to help countries of South Eastern Europe to recover and achieve real progress on their way to economic and social stability, democracy and European integration. In this respect the project delivered a substantial contribution to the implementation of the Stability Pact and the Stabilization and Association Process of the European Union (REC 2006).

4 Conclusion

The process of cooperation for the management of shared natural resources between Albania and Montenegro was based on willingness, common understanding and the commitment of both Albanian – Montenegrin authorities and key actors involved in the process. In this context, the following conclusions and recommendations might be used as an entry point and replicable model for similar regions at the international level:

- One can build joint responsibility for trans-boundary natural resources only by mobilizing an as wide range of players as possible, which on return means that a merely flexible and process-oriented approach is required to maintain their active participation.
- When complex issues like the management of shared natural resources are addressed, capacity building becomes a similarly complex and permanent process entailing a wide range of methods and targeting not only institutions, but also individuals and entire communities.
- Constant trust and confidence in the fairness of dealings within the project – a precondition for maintaining an authentic trans-boundary dialogue on shared natural resources – requires transparency and openness, and a systematic approach to maintaining these.
- Trans-boundary cooperation mechanisms can become sustainable on the long term, when they turn to be an internal value of local actors, local institutions are mandated through bottom-up processes to use these mechanisms, and local actors receive the possibility to address the future of their achievements and prepare for it.
- Synergies among actions happening inside or outside the project or at different levels – local, national, international – are just the same important as the actions themselves. However, exploring them needs an extra effort and a project team that is able to operate with the similar efficiency at the local, national and international level.
- Raising issues concerning the management of shared natural resources can mobilize local authorities, give them new opportunities to learn, create new mandates to facilitate trans-boundary processes, and underline their importance concerning local developments.
- When formerly separated communities are to (re)start their cooperation, addressing a neutral topic – such as the joint management of shared natural resources – ensuring a neutral platform for discussions and providing an independent facilitator can become key preconditions for success.

References

AKM/NEA (ed) (1999) Shqipëria: Konventa mbi Larminë Biologjike. Strategjia dhe Plani i Veprimit për Biodiversitetin. Agjencia Kombëtare e Mjedisit. Tiranë. f. 1–100. http://planet.uwc.ac.za/nisl/Biodiversity/pdf/al-nbsap-01-en.pdf

Anonymous (2006) WWD ceremonies for the designation of Lake Shkodra. The Ramsar Convention on Wetlands. World Wetlands Day 2006 in Albania. http://www.ramsar.org/wwd/6/wwd2006_rpts_albania01.htm

Bejko D (2006) Vështrim ekologjik i pellgut të Liqenit të Shkodrës. Mikrotezë. FShN, UT, Tiranë. f. 1–110

Bejko D, Bushati B (2008) Trans-boundary cooperation through the management of shared natural resources: the Case of the Shkoder/Skadar Lake. Shkoder, Albania, pp 1–4. http://www.inweb.gr/twm4/abs/BEJKO%20Djana.pdf

Dhora Dh (2005) Liqeni i Shkodrës. University Shkodrës 'Luigj Gurakuqi', Shkodër, pp 1–252

Dhora Dh, Sokoli F (2000) Liqeni i Shkodrës – Biodiversiteti. ShRMMNSh, UNDP, GEF/SGP. Shkodër: 10–13, 14–23, 58–74

Ervin J (2003) WWF: Rapid Assessment and Prioritization of Protected Area Management (RAPPAM) Methodology. WWF Gland, Switzerland, pp 1–52

Kashta L, Dhora Dh, Rakaj M (1997): Lista e bimëve dhe kafshëve të Liqenit të Shkodrës – Pjesa shqiptare. (Checklist of plants and animals of Shkodra Lake – Albanian part) – Universiteti i Shkodrës "Luigj Gurakuqi", Sektori i Bioekologjise për Liqenin e Shkodrës, Shkodra, pp 1–24

REC & SDC (ed) (2001) Biodiversity database of the Shkodra/Skadar lake (Checklist of species). The Regional Environmental Center for Central and Eastern Europe (REC) Swiss Agency for Development and Cooperation (SDC). Project: Promotion of networks and exchanges in the countries of the South Eastern Europe. Activity 3. 2. 1. 1–52. http://archive.rec.org/REC/Programs/REREP/Biodiversity/docs/ShkoderBiodiversityDB.pdf

The Regional Environmental Center for Central and Eastern Europe (REC) (2006) Opening doors – opening minds, Hungary, pp 1–46

The Regional Environmental Center for Central and Eastern Europe (REC) (2007) Trans-boundary cooperation through the management of shared natural resources, Hungary, pp 1–93. ISBN: 978-963-9638-16-7

Ziu T, Bejko D (2004) Liqeni i Shkodrës dhe rrethinat – Natyrë me vlera të shumanshme/Shkodra lake and suburbs – Landscape with many values. REC ed., Shkodra, pp 1–164

Environmental Impact Assessment in a Trans-Boundary Context in Montenegro

Lazarela Kalezić

Abstract In this paper it is given an overview of the legislation concerning issues of cross-border impact on the environment in Montenegro, which appropriately regulates the issue and contribute to creating the conditions for the implementation of this management mechanism. In the process of harmonizing national legislation with the *acquis* of the European Union – (*Acquis Communautaire*), a series of new laws were adopted, namely: the Law on Strategic Environmental Impact Assessment (2005), the Law on Environmental Impact Assessment (2005), the Law on accession to the Convention on the Assessment of Environmental Impact in a Trans-boundary Context – Espoo Convention (2009). In the second part of the paper, specific examples are given regarding the application of the EIA procedure in a trans-boundary context in Montenegro. Such examples relate to the project for the construction of the hydroelectric power plant (HPP) "Buk-Bijela", to the project HPP "Ašta" and to the current and planned activities regarding the project of multi-purpose HPP on the Morača River.

Keywords EIA • Espoo convention • Trans-boundary environmental impact assessment • Montenegro

1 Introduction

The Accession to the European Union (EU) is a long-term goal within the overall development strategy of Montenegro. Therefore, when it comes to environmental protection, it is very important to ensure that the Montenegrin policy and legislation is in accord with existing European policies. Several key laws harmonized with the EU directives and accession to relevant international treaties, which are significant

---

L. Kalezić (✉)
Secretariat for Spatial Planning and Environmental Protection, Municipality of Podgorica, Njegoševa, 13, 20 000 Podgorica, Montenegro
e-mail: lkalezic@yahoo.com

for environmental protection and related to EIA in trans-boundary context are considered in this paper. Then, three cases are analysed which fully reflect problems in practical implementation of the EIA legislation in a trans-boundary context, from the standpoint of Montenegro.

2 Montenegrin Legislation

2.1 Introductory Note

In the process of harmonizing the national legislation with the *Acquis Communautaire*, a series of new laws have been adopted, which provide a legal framework in the field of environmental protection and also an efficient use of natural resources and contribute to strengthening the responsibility for our own development. Novelties that covered the given legislation implied the creation of procedural mechanisms for greater public involvement in decision-making process and also the question of cross-border consultations.

The law on Strategic Environmental Impact Assessment (SEA) and the Law on the Environmental Impact Assessment (EIA) were adopted in 2005 as the result of realization of the Project "Development of environmental protection legislation in Serbia and Montenegro" which was made on the basis of an agreement between the Government of the Republic of Finland and the Government of the Federal Republic of Yugoslavia signed in Belgrade on 23 July 2002.

The implementation of these laws began in Montenegro in January 2008 due to the estimation that there were not sufficient relevant capacities on national and local level to implement them and also the need for training and education in order to overcome the problem.

2.2 Strategic Environmental Impact Assessment

The Law on Strategic Environmental Impact Assessment has transposed to a greater extent the provisions of the EC Directive 2001/42 on assessment of the effects of certain plans and programs on the environment, the EC Directive 2003/35 of 26 May 2003 which allows public participation in the drafting of certain plans and programs related to environment and the EC 2003/4 of 28 January 2003 on public access to information about the environment. Moreover, during the drafting of the respective Montenegrin legislation, the EC Habitat Directive 1992/43 was also taken into account.

The Law defines conditions, manner and procedure for the impact assessment of certain plans and programs on the environment, by integrating the basic principles of environmental protection in the process of preparation, adoption and implementation of plans and programs that have a significant impact on the environment [2]. Objectives of the strategic impact assessment are to ensure the inclusion of issues related to environment and health, in plans and programs development, by establishing clear, transparent and effective procedures for strategic assessment, public participation, by providing sustainable development and improving the level of protection of health and environment. The basic principles of strategic impact assessment are the principle of sustainable development, the principle of integration, the precautionary principle, the principle of hierarchy and coordination and the principle of the public consultation.

The competent authority for the implementation of the strategic impact assessment procedure is a state government body responsible for plans and programs that are adopted by authority at the state level, or a body of local government – for plans and programs that have been adopted by an authority at the local level.

Strategic impact assessment is mandatory for all plans and programs that are prepared in the field of spatial and urban planning or land use, agriculture, forestry, fishing, hunting, energy, industry, including the mining, transport, tourism, telecommunications, regional development, waste management, water management, marine area management, and which provide the framework for future development projects that are subjects to strategic environmental impact assessment in accordance with the special act, and also for those plans and programs that, given the area in which they are implemented, may affect protected areas, natural habitats and conservation of wild flora and fauna.

The Law defines the procedure for exchanging information concerning potential cross-border impacts. The Law defines that the exchange of information on cross-border environmental impact on the plan or program will be provided by the state government body in charge of environmental protection. If the implementation of a plan or program may have a significant negative impact on the environment of another state or if the state whose environment may be significantly affected so requests, the state administrative body in charge of environmental protection, in the participation process of the interested organizations and the public, must provide to another state the following information:

1. Description of the plan and program, together with all available information about their possible impact.
2. Nature of decisions that can be taken.
3. Period in which the other country may notify its intention to participate in decision-making process.

The state administrative body in charge of environmental protection will inform the other state, which was consulted in the decision-making, about the decision on granting consent to the strategic impact assessment report, by submitting information on:

1. Content of decision on granting consent.
2. Way that report on strategic impact assessment was made and the opinion obtained in the process of making.

3. Results of the consultation and the reasons on which decision on granting consent is based.
4. Measures in the field of monitoring plans or programs.

The state administrative body in charge of environmental protection shall notify interested organizations and the public about received information on cross-border impacts of the proposed plan or program of another state, in the manner established by law.

Pursuant to the above, the body responsible for preparing the plan or program informs the public on the manner and terms of insight into the content of the strategic impact assessment report and submission of opinions, as well as on the time and place of holding debate.

Public debate may be held not earlier than thirty days after notification to the interested public.

The state administrative body in charge of environmental protection has to take into account the results of the consultations and the obtained opinions of interested bodies, organizations and the public when giving opinions to the competent authority of another state.

2.3 Environmental Impact Assessment

The Law on EIA has transposed the provisions of EC Directives 85/337 and 97/11/EC on the impact assessment of certain public and private projects, in conjunction with those of the directives related to public participation and the right to judicial protection. According to a report of the European Commission, based on indicators from the project "Monitoring of the progress of potential state pre-candidates and the former Yugoslav Republic of Macedonia", the assessment of transposition of the EU requirements in this case was at level of 98%.

This law regulates the impact assessment procedures for projects that can have significant impacts on the environment, the content of the EIA study, the participation of interested bodies and organizations and the public, the process of evaluation and issuance of consent, the cross-border information for projects that can have significant impacts on the environment of another state, the supervision and other issues of importance for environmental impact assessment [1]. Impact assessment shall be conducted for projects from industry, mining, energy, transport, tourism, agriculture, forestry, water management, community activities, as well as for all projects that are planned in a protected natural area and in vicinity of protected immovable cultural property.

Pursuant to the provisions of Article 4 of the Law, the competent authority for the implementation of impact assessment procedure is the competent body of state or local government in charge of environmental protection.

The Government of Montenegro adopted the Decree on the projects for which an impact assessment is done and which determines the list of projects or activities for which the impact assessment procedure is mandatory, as well as list of projects for which the impact assessment procedure may be required and which shall be decided by the competent authority on a case by case basis. The impact assessment procedure consists of the following stages:

1. Decision on the need for impact assessment.
2. Determination of the scope and content of the impact assessment.
3. Decision on granting consent to the impact assessment.

This Law also has clearly defined powers and appropriate procedures for preparation of project documentation, establishing special arrangements for public participation beyond the boundaries of competencies and other related elements of international cooperation. According to the provisions of the Law, if the proposed project may have an impact across the state border, then Montenegro, as the country of origin and the project manager, will ensure that affected state and its citizens in the area of likely impact have the opportunity to participate in the EIA procedures and that appropriate information on assessment at the earliest stage of the process will be provided.

A manual for the application of the Law on environmental impact assessment gives guidelines to create opportunities for establishing a joint body for the exchange of information and organizing public participation in the area that will likely to be affected. Everyone should get certain information and be able to participate in the assessment and give comments on the results. Also, as a possibility, and in order to prevent complication, mechanisms of harmonization of basic information, approach and methodology of assessment among neighbouring countries should be considered, so as to ensure the coincidence between the results obtained. It also points to the need to consider this issue when taking into account the UN Convention on the impact assessment in the trans-boundary context – Espoo, which provides a comprehensive framework for the implementation of activities related to cross-border impact. First of all, same basic start-up procedures for the activities, which are included in the list of activities for which a cross-border impact assessment is required pursuant to the Convention [6].

In connection with cross-border impact, there are two possible cases: one when the project that may have negative impact is realized on the territory of our country, and the other when a certain project is implemented in the territory of the other neighbouring state but can have negative consequence for the environment of our country. In this connection, there are different possible scenarios. If the project may have a significant impact on the environment of another state or if the country whose environment may be significantly affected so requests, the state administrative body in charge of environmental protection of Montenegro, as soon as possible and no later than the deadlines for informing their public, provides to other state information on:

1. The project together with all available data about its possible impact.
2. The nature of decisions that can be taken.
3. The period in which another country may announce its intention to participate in the process of impact assessment.

Further, the competent authority is obliged to notify the state that participated in the process of impact assessment on a decision that was taken as requested, submitting notice of:

1. Content of the decision and certain conditions if those have been determined.
2. Reasons on which the decision is based, including the reasons for accepting or rejecting the submitted objections, suggestions and opinions of interested bodies and organizations and the public.
3. Most important measures that the project manager must take in order to eliminate, prevent, mitigate or remedy adverse consequences.

The procedure that shall be used if the competent authority has received notice of cross-border impacts of the proposed project in another country means public notification. Once having obtained the opinion of the interested public, the state administrative body in charge of environmental protection shall take it into account when giving opinions to the competent authority of another state.

Notification and consultation with other countries on possible trans-boundary impacts shall be done on the principle of reciprocity, according to the concluded international agreements.

3 Relevant International Treaties

3.1 The Espoo Convention

At the end of 2008 the Parliament of Montenegro adopted the Law on accession to the Espoo Convention. The Convention has two amendments and Montenegro, with the ratification of the Convention has also ratified the First and the Second Amendment. This law allows the implementation of the provisions of the Convention, which Montenegro recognized as an extremely important international act in the treatment of environmental problems, as well as in international cooperation in this field [3].

Also, as a significant component of the Convention, there is the possibility for Montenegro to establish, through the acceptance and application of impact assessment regulations, consistent criteria with the requirements in terms of preserving and improving the quality of the environment. Namely, the need for confirmation of the Convention is conditioned by the fact that often the interests of holders of the project are opposed to social interests in protecting and improving the environment. To reconcile competing interests in protecting and improving the environment is necessary to consistently implement procedure of assessment, the manner of its verification and to ensure public access to the document produced.

Therefore, taking into account the situation in developed countries and the need for integration of Montenegro into the modern trends in the field of environmental protection, especially with regard to the realization of cooperation with neighbouring countries through the exchange of information related to activities undertaken in the field of environmental impact assessment in cross-border context, it was necessary to properly implement and institutionalize the environmental impact assessment procedure in Montenegro.

In fact, the ratification of the Convention may provide a political and economic incentive for Montenegro, for obtaining funds from international organizations, for promoting investment projects for building new and reconstruction of existing facilities which would contribute to the overall efforts of environmental protection. This is because some of the international financial institutions (World Bank, European Investment Bank, European Bank for Reconstruction and Development) require the respect of the environmental impact assessment legislation as a condition for giving credits for investment projects.

3.2 The Protocol on Strategic Environmental Impact Assessment

On the fifth Ministerial Conference "Environment for Europe" in Kiev, Ukraine held on 21–23 May 2003 representatives of the state delegation of the Serbia and Montenegro signed the Protocol on Strategic Environmental Impact Assessment. The Law on Ratification of the Protocol was adopted on 15 July 2009 and published in the "O.G. No. 02/09" [5].

3.3 The Law on Ratification of the Multilateral Agreement of SEE Countries on the Implementation of the Convention on the Environmental Impact Assessment in a Trans-Boundary Context

With a financial support from the Government of Switzerland, the Swiss Agency for Development and Cooperation – SDC and technical assistance of Secretariat of the Espoo Convention, the project of making "Multilateral agreement between countries of South-Eastern Europe" was realized. In the period 19–21 May 2008 in Bucharest, Romania, the Fourth Meeting of States Parties to the Espoo was held and the SEE countries signed a multilateral agreement aimed at better implementing EIA in a trans-boundary context.

This agreement establishes obligations of the Parties to take all the necessary legal, administrative and other measures to enforce the provisions of the Espoo Convention and the obligations to adopt criteria for identifying harmful trans-boundary impacts and ways of action in the drafting of specific strategic, planning and development documents [4].

Through the ratification of this agreement, Montenegro extends the legislative framework which provides for the establishment of relations in the field of environment in a more complete way. The application of the provisions of this Agreement and the Convention creates the preconditions for the prevention of negative environmental impacts of projects that intend to be realized by the neighbouring countries in the region. By signing this agreement, Montenegro also expressed willingness to participate in international actions aimed at a better coordination of the legal and regulatory frameworks of the member states with the Convention, to a higher level of exchange of experience that would lead to better implementation of the Convention, helping to further defining implementation of the Convention for the concrete conditions (natural, economic, social) by detailed specifications through bilateral or multilateral agreements that would state application of the Convention between specific countries. Similarly, the signing of a multilateral agreement creates the preconditions for the improvement of cooperation in the field of environmental impact assessment in a trans-boundary context and the development of cooperation with neighbouring countries and countries in the region. This agreement emphasizes the need to improve monitoring and reporting on environmental condition.

4 Examples of Practical Implementation of the EIA Legal Frameworks

As examples of the EIA in a Trans-boundary Context in the following part of the paper, the situation regarding the project of construction of hydroelectric power plant (HPP) "Buk-Bijela", the project HPP "Ašta" and the current and planned activities related to project of multi-purpose HPPs on the Morača River will be analyzed. All the three projects in various ways concern precisely cross-border issues. Those cases are not examples of completely typical EIA procedure in transboundary contexts, but show the current situation in Montenegro on this issue and try to provide some inputs on how to improve the situation.

4.1 The Construction of the Hydropower Plant "Buk-Bijela"

4.1.1 Introductory Notes

HPP "Buk-Bijela" was planned to be built on the Drina River in the Bosnia and Herzegovina, Republic of Srpska, but a part of the accumulation was supposed also to impact 10 km of the Piva River and 12 km of the Tara River, the later one being on the UNESCO List of World Biosphere Reserves under the "Man and Biosphere" Program (UNESCO MAB Program) since 1977. At the same time, that part of the Tara River Canyon, which would be submerged and become the accumulation of HPP "Buk-Bijela", is located nearby the borders of National Park "Durmitor", which has been on the UNESCO List of world natural heritage since 1980. The planned project provoked a popular reaction in Montenegro against the decision on construction of HPP "Buk-Bijela".

Basic technical and energy parameters of HPP "Buk-Bijela" refer to the production of 1,345 GW per year. Height of the dam is planned at 125.6 m, the surface catchment area of 4.033 km^2. The project would be realized in an arrangement in which the concessionaire would take over the construction, financing, operation and maintenance of the hydropower facility. The concessionaire would take over the exploitation of the plant in a certain time period during which the Electro industries of Montenegro and the Republic of Srpska would buy the whole or part of the electricity produced at the tariff established in advance – two thirds to Republic of Srpska and one third to Electric Power Industry of Montenegro. The concession would be granted for 30 years. After the expiration of 30 years, the ownership of HPP "Buk-Bijela" would be divided by already established key – two-thirds to Republic of Srpska and one-third to Montenegro.

It is worth noting that the construction of HPP "Buk-Bijela", about 10 km downstream from Foca (Srbinje) (Republic of Srpska), began in the 1970s of last century, and in 1974 the works were stopped and the building preserved because of the enormous resistance of the domestic and world expert public (over 500 scientists from all around the world and also the World Geographers Congress in Edinburgh 1984 were against this project because of the catastrophic consequences on the natural environment).

For the purposes of the planned project, in 2000, environmental impact assessment was done.

4.1.2 Chronological Overview of Relevant Activities

In April 2004, the Government of Montenegro considered and adopted the Agreement on Cooperation between the Republic of Srpska and the Republic of Montenegro on the implementation of the construction and joint use of a hydroelectric power plant "Buk-Bijela". The Agreement was supposed to be signed by the Prime Ministers or be ratified by the Parliaments of Montenegro and the Republic of Srpska. However, the signing did not occur, because certain conditions of the Agreement were not met [9].

In April and May 2004, the Montenegrin NGO sector launched a civil initiative, which expressed opposition and protest against the realization of the project HPP "Buk-Bijela". The public campaign "I do not want a puddle, I want Tara" has been developed as a strong response to the announcement of senior officials of the Government of Montenegro and the Republic of Srpska that building of HPP "Buk-Bijela" affect the Drina River, which would, because of the scope of its accumulation, submerge the lower part of the colourful Tara River Canyon, the parts of the Piva River Canyon and the Sutjeska River Canyon, as well as because of disabled access to information about planned project. During the campaign, the demand was announced "for all citizens of Montenegro to be free to declare them on this issue". In a statement, the public was "warned of the disastrous economic and environmental consequences of implementation of this project" [7, 13].

In this regard, the Ministry of Environment and Spatial Planning of Montenegro, through press releases and press statements on this issue expressed its intention to conduct expert assessment of existing Study on environmental impact assessment of construction of HPP "Buk-Bijela", with the application of international standards. Pursuant to such intention, the Montenegrin Ministry addressed in June 2004 the Minister for Spatial Planning, Engineering and Ecology of Srpska Republic with the request for submission of the Study on environmental impact assessment for the construction of HPP "Buk-Bijela". However, no answer was received.

In the meantime, activities of NGOs had been intensified through the signature of the petition against the project as well as through securing international support from a large number of environmental NGOs from the region and from all over Europe [7, 13]. Considering the overall situation, in order to provide a comprehensive analysis of the problem, the Montenegrin Ministry invited in 2004 UNESCO World Heritage Center to propose experts that would be engaged in professional discussions about the Study on environmental impact assessment. The answer provided by UNESCO expressed concern that the planned activities could jeopardize the National Park "Durmitor" as an area that is on the UNESCO list of World Cultural and Natural Heritage and expressed willingness to provide a review of the Study on environmental impact assessment by experts of the International Union for Conservation of Nature (IUCN). The problem was also presented to the Department of Environmental Sciences of UNESCO Office in Venice, which supported the need for drafting a new Study on environmental impact assessment and implementation of quality public and expert discussions, with the participation of representatives of the Department.

Considering the undertaken activities, the competent Ministry of the Republic of Srpska was addressed, with regard to the development of a new Study on environmental impact assessment and a proposal concerning a joint invitation of the World Bank their experts to create the new Study on environmental impact assessment.

Although there were no exact indicators, it could be assumed that existing environmental Study 2000 did not properly address all the environmental impacts of the project. Unfortunately, despite all the efforts, the existing Study from 2000 was not obtained from the Republic of Srpska, which also did not accept proposal for invitation of the World Bank for making a new Study.

In October 2004, a mission of UNESCO was invited to visit Montenegro and to conduct on-site technical elaboration of the planned interventions and assessment of the likely impacts on the environment. Considering that it would be good that UNESCO experts at this stage present their views and assessments, they were submitted on the environmental impact study of 2000, that was obtained in unofficial way.

In the meantime, NGOs continued with their campaign through a series of protests and performances. As the most important should be highlighted activity of NGO "Bridge (Most)" which, with the team of legal experts, drafted the Declaration on the Protection of the Tara River, which was submitted for approval to the Assembly of Montenegro. The submission of this document was preceded by the collection of the necessary 6,000 signatures of citizens of Montenegro, to launch a parliamentary debate on the adoption of the Declaration. The action "Montenegro for Tara", which was held in September 2004, was supported by more than 20

NGOs and was gathering signatures to run parliamentary debate on the Declaration, in all towns of Montenegro. The declaration was supported by more than 10,000 citizens. The action was accompanied also by the international public media. The Parliament of Montenegro adopted, with a majority vote on 14 December 2004 the Declaration on the Protection of the Tara River. Such a Declaration prevents the construction of facilities which would violate the Tara River Canyon.

In January 2005, an expert mission of UNECO/IUCN was sent to Montenegro. On that occasion, the Mission visited the area of the National Park "Durmitor", the Tara River and also the municipality of Foča (Republic of Srpska). A roundtable on "Protection and valorization of the Tara River" was organized by the Montenegrin authorities at which participated representatives of the Government of Montenegro, relevant institutions, the Republic of Srpska Government, NGOs sector and members of the mission of the UNESCO/IUCN [14].

Expert mission UNESCO/IUCN in March 2005 submitted its report to the Ministry, which specified that a decision will be made about inserting the NP "Durmitor" on the UNESCO list of World Heritage in danger, if the implementation of activities on the construction of HPP "Buk-Bijela" would continue. The mission pointed out that the proposed project should be evaluated in a trans-boundary context, by the interested States, as well as UNESCO, Bosnia and Herzegovina and Serbia and Montenegro. The mission particularly stressed that both Montenegro and BiH are Parties to the Convention on the World heritage, and that Article 6.3 of the Convention specifically refers to this specific case: "Each State Member to this Convention is obliged not to take any deliberate actions that would directly or indirectly could damage cultural and natural heritage ... located on the territory of other State Member to this Convention" [9].

Furthermore, the Mission noted in the report that the Ecological study on environmental impact assessment (2000), in the form in which it was officially presented to the Mission, could not be accepted as an appropriate Environmental impact assessment study. The Mission stressed that it seemed that the study was a copy of previous assessments that have been done for previous construction projects of hydroelectric power plants (1970 and 1980). All of these previous projects, in the same location, were stopped due to strong public opposition and other circumstances. The Study repeated mistakes from previous elaborates on environmental impact assessment, which shows a lack of scientific basis. As it was pointed out on the round table held in Žabljak, even the term of the Tara River Basin was not specified and overlapped with the Tara Mountain and the National Park "Tara". Therefore, any impact on flora and fauna that was mentioned in this part of the study could not apply to the area that was planned to be submerged as a result of the construction of HPP "Buk-Bijela". Data regarding the length of the Tara River and the areas that should be submerged also vary and range from 12–18 km of the canyon (from the location of the dam) and there is concern that these mismatched can be greater. During the examination of maps, the Mission noted that the area of the Tara River – as a natural monument would be submerged, as it is near the north-eastern border of the National Park "Durmitor". Furthermore, part of the Tara River Basin as Biosphere Reserve (which coincides with the border part) to the place where the river borders with the Piva River, on south, would be flooded.

In response to that, the Government of Montenegro urged the Republic of Sprska to comply with the recommendation of UNESCO and to stop further activities concerning the possible construction of HPP "Buk-Bijela". In addition, it asked for a study on environment impact assessment, which would be harmonized with EU standards and that would determine the impact that the construction of HPP "Buk-Bijela" could have on the Tara River and the NP "Durmitor" [9, 12]; At the 29th session of the Committee for the World Heritage of UNESCO, held in Durban, South Africa from 10 to 17 July 2005. Within point 7, subpoint 21, it was adopted the Decision on NP "Durmitor" from which, after submitting the letter by the Ministry of Environment and Physical Planning in which was expressed the Montenegro Governments position on stopping further construction activities on HPP "Buk-Bijela", was deleted the article that predicted putting Durmitor on the UNESCO List of World Heritage in danger [9].

4.2 The Construction of the Hydropower Plant "Ašta" on the Drim River, Albania

In September 2008, the company *Verbund* from Vienna received a 35-year lasting concession for construction of the Hydroelectric Power Plant Ašta. Its annual production capacity is designed at 230 million kWh, and it is envisaged that the construction will last 40 months and will be completed in 2012 [11, 15].

Montenegro has not been officially informed by the Government of Albania on the activities undertaken on the realization of the construction of HPP "Ašta", as it is clearly defined in the provisions of the Espoo Convention. On the current case, the Government and the public of Montenegro were informed by the organization REC Albania (Regional Environmental Center), which organized the public debate in Shkodra on the topic "Environmental Impact Assessment of the construction of hydroelectric power plant Ašta on the Drim River" in cooperation with NGOs from Albania and Montenegro.

A public debate was attended by representatives of the Agency for Environmental Protection of Montenegro invited by a NGO from Montenegro. In the introductory part, the environmental impact assessment study on the construction of HPP "Ašta" on the Drim River was introduced to participants. The study, to some extent, gave a description of the impacts that this project could have on the environment at the site of the construction and completely ignored the review and evaluation of long-term impacts that this project could have on the Skadar/Shkodra Lake and the Bojana River Delta. The study even concluded that the project shall not impact protected areas [8].

Connoisseurs of the issues related to the natural situation in these areas expressed their criticism in regards of the lack of consideration of the possible environmental impacts of this project. They also expressed dissatisfaction over the fact that the competent institutions in Montenegro were not officially informed on the project by the competent institutions of the Republic of Albania. At the end of the public discussion, it was concluded by the organizer that the report from the public debate, with all the opinions on open issues and objections, will be submitted to the competent authorities in the Republic of Albania and to the working group that worked on the Study on environmental impact assessment on the construction of HPP "Ašta" [9, 10].

Both Montenegro and the Republic of Albania are parties of the Espoo Convention. Pursuant to the above, the Ministry of Physical Planning and Environment and the Ministry of Foreign Affairs sent a diplomatic note to the Ministry of Foreign Affairs of the Republic of Albania, requesting information on the on-going project.

4.3 The Detailed Spatial Plan of Multi-Purpose Accumulations on the Morača River, Montenegro[1]

The Government of Montenegro adopted the Decision on the preparation of the Detailed spatial plan for the area of multi-purpose accumulations on the Morača River in October 2008. Detailed Spatial Plan (DSP) for the Hydropower Plant on the Morača River is part of the ambitious Strategy for Energy Development (SRE) and the Action Plan of Montenegro for the period by 2025, representing a step forward towards the reduction of dependence on imported electricity, primarily through the creation of stable conditions for investments in research/exploitation and development of new energy facilities (especially in the already explored sites with unexploited hydropower potentials), as well as investments in other energy infrastructure. In Montenegro, there is a significant deficit of electricity and unused hydropower potential of Montenegro is also a much needed resource of foreign trade exchange.

Along with this, the Decision on making Strategic environmental impact assessment (SEA) for the Plan was taken. In this regard, the Ministry of Economic Development of Montenegro (MEM) addressed the Government of the Kingdom of Norway for technical assistance in the preparation of SEA. The estimated time required for the construction of four HPP on the Morača River is 6 years. At the time of writing this paper, SEA was on the pre-draft level. SEA should be focused on specific aspects of the DSP, and it is expected to contain assessments of the most significant effects on the environment and a reference to the proposed solutions, their impact on the environmental, social and economic situation, as well as proposal for appropriate actions to avoid, reduce and/or mitigate the negative impacts and maximize the positive ones. Due to the fact that the potential downstream effects of the DSP will include cross-border issues, SEA has recognized the importance of Espoo Convention as an important and relevant document, with the UNECE Protocol on SEA (Kiev Protocol) which is also relevant in this case.

The Skadar/Shkodra Lake has been set aside as a separate landscape unit due to the distinctive appearance to the environment and the extraordinary harmony of natural and cultural heritage. The area is protected by law as a national park, located on the Ramsar List of Wetlands of International Importance.

---

[1] Author is Member of Commission for monitoring the project documentation of multipurpose accumulation on the Morača river.

It should be noted that the Montenegrin Government has an active dialogue with Albania on the issues related to the Skadar/Shkodra Lake. The dialogue between two Governments concerning the environment of the Lake began in 1994 with the Protocol on Cooperation signed between the University of Shkodra and the University of Montenegro for the implementation of scientific research on the Skadar/Shkodra Lake. In May 2003 a Memorandum of Understanding (MoU) for the protection and sustainable development of the Skadar Lake was formally signed by the Ministries of Environment of Albania and Montenegro.

In accordance with the SEA Law, the Government of Montenegro will adopt the Draft Plan and Report on Strategic Assessment and program of public hearings that will last 30 days. Accordingly, official communication with the Republic of Albania will be established with the submission of relevant information and documents and will enable a meaningful participation in the SEA procedure.

5 Conclusion

Having in mind what was said above, it can be concluded that, in terms of legislation, the harmonization of Montenegrin EIA procedure with the EU requirements has been achieved with regards to cross-border projects. However, the analysis of concrete cases shows that some problems exist for environmental impact assessment in a trans-boundary context. The cause could be found in the absence of experience of competent authorities of all interested parties and in the lack of sufficient precise guidelines for the implementation of a trans-boundary EIA procedure. However, it should be emphasized that at the time of the events related to the case HPP "Buk-Bijela", it did not exist any corresponding legislation that should define this issue. However, for the time ahead, in front of Montenegrin institutions are the challenges posed by the running case of HPP "Ašta" and the activities under the project of building multi-purposed accumulation on the Morača River.

References

1. Law on Environmental Impact Assessment ("Official Gazette of Montenegro", No 80/05)
2. Law on Strategic Environmental Impact Assessment ("Official Gazette of Montenegro", No 80/05)
3. Law on ratification of the Convention on the Environmental Impact Assessment in Trans-boundary context – Espoo ("Official Gazette of Montenegro – Multilateral Agreements", No 08/08)
4. Law on ratification of Multilateral agreement of SEE countries on the implementation of the Convention on the environmental impact assessment in a Trans-boundary Context ("Official Gazette of Montenegro – Multilateral Agreements", No 02/09)
5. Law on ratification of the Protocol on Strategic Environmental Impact Assessment in a Trans-boundary Context ("Official Gazette of Montenegro – Multilateral Agreements", No 02/09)

6. Kalezić L, Knežević B, Bušković V (2009) Manual for the application of the Law on Environmental Impact Assessment. Regional Environmental Center – Office for Central and Eastern Europe (REC), Podgorica

Internet Resources

7. www.durmitor.com
8. www.epa.org.me
9. www.gov.me/files
10. www.gov.me/minurpizzs
11. www.monitor.co.me
12. www.mup.vlada.cg.yu
13. www.ngo-most.org
14. www.visit-montenegro.com
15. www.visit-ulcinj.com

The Framework Agreement on the Sava River Basin (FASRB)

Jasnica Klara Matic

Abstract The Framework Agreement on the Sava River Basin—FASRB (2002) is the first regional (river basin) agreement, except the Dayton Peace Agreement (1995), concluded by the countries originated by the decomposed Socialist Federal Republic of Yugoslavia (SFRY). The conclusion of the FASRB was strongly supported by the international comunity, acting intensively in the Balkan region after the wars in the nineties. This Agreement is an example of positive development in managing of natural (i.e. water) resources in a post-war period.

The different phases and aspects of drafting, negotiation, signature, ratification and implementation of the Framework Agreement on the River Sava Basin are presented here. The paper contains brief data on the geographical, historical and political features of the Sava River Basin, on the former Yugoslavia disintegration, on the Stability Pact for South Eastern Europe's role in the proces of drafting the Agreement as well as on the Sava Initiative description. The paper also contains details on the goals, general principles, areas of cooperation, mechanism of cooperation, monitoring implementation dispute settlement, protocols, the Statute of the International Sava River Basin Commission and the dispute settlement provisions.

A separate part of the paper comprises some details on the current status of cooperation in certain fields covered by the Agreement, including the status of (inland) navigation, with brief references on the FASRB implemenation in the Republic of Croatia. The efforts on the unification of the basic navigation rules, establishment, maintenance and improvement of the uniform waterway marking system as well as on the rehabilitation and development of the Sava River Waterway and development of the River Information Services are presented. The activities concerning water management issues such as the development of the Sava River Basin Management Plan, the establishment of a system of measures, activities, warnings and alarms in case of extraordinary impacts of the water regime caused by pollution or floods, ice occurrence, droughts and water shortages, the establishment

---

J.K. Matic (✉)
EQUILIBRIUM, Environmental Law Association Croatia, Bernarda Vukasa 39/3, Zagreb 10000, Croatia
e-mail: jasnica.klara.matic@zg.t-com.hr

of an integrated information system, as well as provisions regulating dispute settlement are highlighted. Activities on drafting new treaties, in terms of Protocols to the FASRB, are also reviewed.

Keywords FASRB • Sava Commission • Sava River Basin • Inland navigation • Water management

1 Introduction

The Sava River is the third longest, and the largest by the size of discharge, tributary of the Danube River. It runs through four countries from spring in Slovenia, crosses Croatia, forms the border with Bosnia-Herzegovina and terminates in Serbia flowing into the Danube at Belgrade. It is 946 km long and its basin, with an area of 97,713 km^2, covers the bigger parts of Slovenia, Croatia, Bosnia and Herzegovina, Serbia and Montenegro, as well as a small part of the territory of Albania. Albania and Montenegro are not parties of the FASRB.

Prior to the disintegration of Yugoslavia, the Sava River was the biggest national river with proper legal and administartive instruments for water management and was navigable. After the disintegration in 1991, it has become an international river. The great political and economic changes that have taken place in the Sava River Basin countries, after the disintegration of Yugoslavia, have initiated engagement of the Stability Pact for South Eastern Europe (Stability Pact) for promoting a coordinated effort in defining and organising the best utilization of the Sava Basin water and related resources. The Stability Pact Initiative (the Sava River Initiative) was aimed at establishing and developing an internationally recognized partnership among the Sava River Basin countries based on the models set up for other international rivers.

The Sava Initiative was launched by the Letter of Intent Concerning the International Sava River Basin Commission Initiative (Letter of Intent), signed by the Ministers of Foreign Affairs of Croatia, the Federal Republic of Yugoslavia and Slovenia and the Minister for Civil Affairs and Communications of Bosnia and Herzegovina, in Sarajevo on 29 November 2001. By signing the Letter of Intent, the four countries committed themselves to prepare the draft of a framework agreement based on the existing principles of the law of international water resources which, once ratified, would serve as the basis for establishing an international Sava River Basin Commission. They also accepted to prepare and start-up the implementation of the first Sava Commission Short and Medium Term Master Plan aiming at rehabilitating and developing the navigation of Sava River Basin and its main tributaries Drina and Una, at ensuring the sustainable development, utilization, preservation and management of the Sava River Basin waters and related resources, at preserving and protecting the environment, the biodiversity and the aquatic condition and at promoting the social and economic welfare of Sava River Basin countries.

Negotiations of the Agreement started on March 12, 2002 in Brčko, Bosnia and Herzegovina and were led and chaired by the Special Coordinators of the Stability Pact. Shortly after the beginning of the negotiations, the Special Coordinators agreed to change the method of drafting work. Instead of joint formulation and consensus of delegations of negotiating States on each provision of the Framework Agreement Draft, they agree to deploy an expert team formed jointly by the negotiating States that would draft the text of the river basin agreement on the basis of strong expert arguments.

The final Framework Agreement Draft was agreed at expert sessions which took place at Novi Sad and Stubičke Toplice. The Framework Agreement on the Sava River Basin (FASRB) was signed on December 3, 2002. In the years which followed, all Parties ratified the FASRB, so it entered into force on December 29, 2004.

2 The Agreement

2.1 Goals

The establishment of an international regime of navigation on the Sava River and its navigable tributaries, as well as the establishment of a sustainable water management, undertaking measures to prevent or limit hazards, to reduce and eliminate harmful consequences, including those from floods, ice hazards, droughts and accidents involving substances hazardous to water, through cooperation of the Sava River Basin countries are defined as the main goals of the FASRB. With an aim to achieve these goals, the Parties are obliged to cooperate in the process for the creation and realization of joint plans and development programs on the Sava River Basin and the harmonization of their legislation with the EU legislation.

2.2 General Principles of Cooperation

The Parties to the FASRB are obliged to cooperate on the principles of their sovereignty, equality and territorial integrity, for mutual benefit and in good faith in achieving the FASRB goals. They accepted to cooperate on the basis of, and in accordance with, the EU Water Framework Directive. In the framework of their cooperation, the Parties committed themselves to exchange data and information on the water regime, the regime of navigation, legislation, organizational structures and administrative and technical practices. The Parties also accepted the obligation to cooperate especially with the International Commission for Protection of the Danube River (ICPDR), the Danube Commission (for navigation), the United Nations Commission for Europe (UNECE) and the Institutions of the European Union. They agreed to appoint organizations, authorities or bodies competent for the realization of the FASRB in the part of the Sava River Basin within their territories.

The principle of reasonable and equitable share of the beneficial uses of the Sava River Basin water resources is the basic principle of water law the Parties accepted to apply. Sharing the beneficial uses of waters resources would be implemented in such a way that reasonable and equitable share is to be determined in the light of relevant factors according to international law.

In conjunction with the reasonable and equitable share of the beneficial uses of waters, another principle was introduced, under the title "No Harm Rule". It determined as an obligation of all Parties to cooperate and take all the appropriate measures for the prevention of causing significant harm to other Party or Parties in utilizing waters of the Sava River Basin in their territories. Formulated in the way as it is, that principle is substantially different from the classical static formulation of the principle "No Harm Rule", which, in practice, was never applied. The FASRB is based on the EU concept of prevention of causing significant harm to neghbouring countries applied to the countries of the Sava River Basin.

The Parties agreed to regulate all issues concerning measures the aim of which is to ensure the integrity of the water regime in the Sava River Basin and to eliminate or reduce the transboundary impacts on the waters of the other Parties, caused by economic or other activities. For that purpose, the FASRB provides for the conclusion of a separate protocol to regulate the procedures for the issuance of water right acts (licenses, permits and confirmations), for construction and use of water installations and for activities that may have a transboundary impact on the integrity of the water regime.

2.3 Scope of Cooperation

The Parties agreed to cooperate on the navigation of the Sava River from Sisak to the mouth of the Danube River and on all navigable parts of the Sava tributaries by merchant vessels. Merchant vessels of any state shall be free to enter the ports on navigable waterways for the purpose of loading or unloading, re-supply or other related operations, taking care to respect the national regulations of the party in whose territory the port is situated as well as the rules passed by the International Sava River Basin Commission. Measures shall be undertaken to maintain the waterways within the territories of the Parties in navigable conditions as well as to improve the conditions of navigation and not to prevent or obstruct navigation.

The determination of the fairway of the Sava River and its navigable tributaries that form the border between Bosnia and Herzegovina and the Republic of Croatia remains exclusively within the jurisdiction of those two states. The determination of the fairway of the Sava River and its navigable tributaries that form the border between Bosnia and Herzegovina and the Federal Republic of Yugoslavia (*i.e.* Serbia) remains exclusively within the jurisdiction of those two states (Article 10.8).

The management of the waters of the Sava River Basin shall be performed in a sustainable manner, including integrated management of the surface and ground water resources, with the aim of ensuring:

- water in sufficient quality and of appropriate quality for the preservation, protection and improvement of aquatic ecosystems (including flora and fauna and eco-systems of natural ponds and wetlands);
- water in sufficient quality and of appropriate quality for navigation and other kinds of utilization;
- protection agains harmful effects from water (flooding, excessive groundwater, erosion and ice hazards);
- resolution of conflict of interest caused by different uses and utilizations;
- effective control of the water regime.

A joint and/or integrated plan on the management of the water resources of the Sava River Basin, was envisaged as a key development instrument. The Parties accepted to cooperate on preparatory activities of such plan and to adopt the plan to the proposal of the Sava Commission.

A coordinated or joint system of measures, activities, warnings and alarms in the Sava River Basin for extraordinary impacts on the water regime was agreed to be established. Extraordinary impacts were determined as sudden and accidental pollution, discharge or artificial accumulations and retentions caused by collapsing or inappropriate handling, flood, ice, drought, water shortage, and obstruction of navigation. When establishing a coordinated or joint system of measures, activities, warnings and alarms in the Sava River Basin for the case of extraordinary impacts, the Parties accepted to act in accord with the activities undertaken in the framework of the Convention for the Protection and Sustainable Use of Danube River and in the scope of the procedures agreed within the ICPDR.

2.4 *Mechanisms of Cooperation*

The institutional mechanism established by the FASRB include the Meeting of the Parties and the International Sava River Basin Commission (Sava Commission). The Meeting of the Parties takes place at least once every 2 years, unless the Parties decide differently, or at the written request of any Party. Its main tasks are to review the work and activities of the Sava Commission, to review and adopt proposals of new protocols and amendments to the FASRB and to undertake any other activity that may be required for the achievement of the purposes of the FASRB. All decisions at the Meeting of the Parties must be made by consensus.

The Sava Commision is a joint institution of the Parties whose main task is to ensure the FASRB implementation. The Sava Commission is authorized to make:

- decisions aimed at providing conditions for safe navigation;
- decisions on the conditions for financing construction of navigable waterways and their maintenance;

- decisions on its own work, budget and procedures; and
- recommendations on all other issues regarding realization of this Agreement.

The Parties to the FASRB agreed to establish the permanent implementation monitoring methodology including timely provision of information to stakeholders and the general public by the authorities responsible for implementation of the FASRB. As of now, it seems that such methodology is not yet in place.

2.5 Dispute Settlement

Disputes about the interpretation or implementation of the FASRB, shall be resolved by negotiation among the Parties. If the concerned Parties are unable to resolve the dispute through negotiation, they have at their disposal other means known in international law, i.e. they may jointly seek good services, mediation or conciliation from a third party, or they may agree to refer the dispute to arbitration, or to address it to the International Court of Justice. The procedure for dispute settlement shall begin upon the request of one of the concerned parties.

If within six months from submitting a request of one of the concerned parties, they are still unable to resolve the dispute, any Party concerned may request the deployment of an independent fact-finding expert committee. The fact finding expert committee would consist of three experts. Each Party in the dispute shall appoint one expert. These two experts shall be selected by common agreement as well as a third expert who is not a citizen of any of concerned Parties and who shall be chairman of the Committee.

The settlement of disputes by arbitration is regulated by Annex II to the FASRB. It regulates in detail:

- initiation of the arbitration and role of the Secretariat of the Sava Commission in the procedure;
- composition of the arbitration tribunal;
- arbitration procedure;
- arbitration decision;
- arbitration expenses.

2.6 Protocols

With an aim to ensure the implementation of the FASRB, which is a framework agreement, setting legal basis and conditions for conclusion of new treaties, the Parties shall conclude additional protocols, regulating transboundary impacts, protection against floods, groundwater, erosion, ice hazards, droughts and water shortages, water utilization, exploitation of stone, gravel and clay, protection and improvement of water quality, protection of aquatic ecosystems, prevention of the water pollution caused by navigation, emergency situations and other issues. The needs of the Parties in different stages of the FASRB implementation shall be the criterion for entering into new treaties. As of now, two protocols were concluded: one on the navigation regime (Kranjaska Gora, 2002; in force) and the other on the prevention of water pollution caused by navigation (Belgrade, 2009; ratified by BiH and Croatia).

2.7 Statute of the Sava Commission

The Statute of the International Sava River Basin Commission (contained in Annex I to the FASRB) regulates in detail:

- composition of the Sava Commission;
- chairmanship of the Sava Commission;
- sessions of the Sava Commission;
- tasks and competencies of the Sava Commission;
- competence for taking decisions and recommendations of the Sava Commission;
- financing the Sava Commission;
- secretariat;
- official Languages of the Sava Commission;
- permanent and Ad hoc Expert Groups;
- reports;
- immunities of personnel.

3 Current Status of Implementation

3.1 Status of Navigation Issues

The work of the Sava Commission in the field of navigation have been as of now focused mainly on the unification of the basic rules regarding navigation, technical requirements and crew members, establishment, maintenance and improvement of the uniform waterway marking system, rehabilitation and development of the navigation on the Sava River Waterway and development of River Information Services. At the same time, the Sava Commission made decisions with an aim to implement the basic navigation regulations and worked on the development of the technical rules for transport of dangerous goods with an aim to improve navigation safety and environmental protection.

Pursuant to Article 10, Para. 1 and 3 of the FASRB and Art. 2, 3 and 8 of the Protocol on the Navigation Regime to the FASRB, the Sava Commission, jointly with the Rhine Commission, Danube Commission, Mosel Commission and the UNECE, worked on the harmonization process between the existing navigation rules on the European level. It also worked on the establishment of criteria for mutual recognition of the boatmaster certificates with the aim to minimize administrative obstacles for the development of the inland navigation.

In the field of technical standards, the Sava Comission developed the Draft Technical Rules for the Vessels on the Sava River Basin on the basis of the EC Directive 2006/87 which prescribes the technical requirements for inland waterway vessels. It also developed the Draft Technical Rules for the Vessels on the Sava River Basin according to the European Agreement concerning the International Carriage of Dangerous Goods by Inland Waterways. The adoption of these rules will be the basis for the implementation of highest technical standards in shipbuilding and the transport of dangerous goods and will surely contribute to the improvement of the navigation safety and environmental protection.

In accordance with Article 10, Paragraph 4 of the FASRB and Article 9 of the Protocol on the Navigation Regime to the FASRB, the Sava Commission is competent for coordinating and supporting of activities concerning the waterway marking and infrastructure rehabilitation. In relation to this, the Commission agreed on the Decision 19/08 on the Adoption of Classification of the Sava River Waterway, the Decision 37/08 on Amendments to the Decision 29/07 on Adoption of the Marking Plan for the Sava River and its Navigable Tributaries for Year 2008 and the Decision 02/09 on Adoption of the Marking Plan for the Sava River and its Navigable Tributaries for Year 2009.

The Detailed Design of the Marking System of the Sava River Waterway in the Bosnia and Herzegovina Marking Sector was prepared in the framework of the Sava Commission.

The basic guidelines for the maintenance of the waterways within the territories of the Parties in navigable condition and specif measures to improve the conditions of navigation and not to prevent or obstruct navigation were provided on the basis of the River Transportation Development Strategy and the Operational Transporatation Programme, in accordance with Article 10, Paragraph 4 of the FASRB.

The reconstruction and development of the Sava River Waterway is one of the most important infrastructure projects on all waterways of the Republic of Croatia. It was elaborated in detail in the Medium Plan Water Ways and Ports Development of the Republic of Croatia.

The development of the River Information Services on all navigable ways of the inland waters has been emphasized as a priority and precondition for safe navigation. Current activities in the Republic of Croatia are directed on covering the whole river water way in the Republic of Croatia by the River Information Services.

In the framework of the River Information Services (RIS) implementation activities, the Sava Commission adopted the Decision 03/09 on adoption of the Vessel Tracking and Tracing Standard and the Decision 04/09 on adoption of the Inland ECDIS Standard. The establishment of the River Information Services System would ensure the supervision of navigation as well as increase the safety of the inland water rivers.

3.2 Status of Water Management Activities

The main activities of the Sava Commission in the field of water management include the development of the Sava River Basin Management Plan, the establishment of a system of measures, activities, warnings and alarms for extraordinary impacts of the water regime caused by pollution, the establishment of a system of measures, activities, warnings and alarms of the water regime such as flood, ice, drought and water shortage as well as the establishment of the Geographical Information System (GIS).

The activities on the development of the Sava River Basin Management Plan (Sava RBM Plan) have been undertaken by the Sava Commission according to the steps agreed in the Structure and the Road Map of the Sava RBM Plan and in the List of Contents of the Sava River Basin Analysis Report, pursuant to Article 12 of the FASRB. Firstly, these activities were focused on the preparation of the Sava River Basin Analysis Report aiming at presenting the overview of the general characteristics of the Sava River Basin and assessing the water quality of surface and groundwater and the water quantity of the Sava River Basin as well as at presenting the monitoring data from transnational monitoring network stations, economic elements and the accompanied GIS maps. Secondly, these activities have to be focused on the preparation of separate annexes of the SRB Analysis Report with an aim to present and assess data on flood management, navigation and hydromorphology defined as issues of special importance for the Sava River Basin.

The actual national structure of the emergency warning system in the Republic of Croatia consits of two subsystems: the international Accident Emergency Warning System (AEWS), related to transboundary impacts caused by accidental pollution, and the national subsystem for accidental water pollution with local impacts only.

According to this, the key activities of the Sava Commission in the field of accident prevention and control include testing and improvement of the AEWS, in accordance with Article 13, Paragraph 1 of the FASRB. These activities are focused on improvements and modification of the integrated system to exchange warning, data and information about the accident and a review of the operative structure and the capacities within relevant bodies.

In the field of protection against water pollution caused by navigation, in addition to the development of the Protocol on this particular issue, the Sava Commission is involved in the project Waste Management for Inland Navigation on the Danube (WANDA), which is aimed at developing proposal for the establishment of a harmonized waste management system for the Danube.

The Sava Commission, jointly with the Danube Commission and ICPDR, guided the process of implementation of the Joint Statement on Guiding Principles for the Development of Inland Navigation and Environmental Protection in the Danube River Basin.

The National Protection and Rescue Directorate, Sector for 112 System, serves as a Communications Unit (CU) for both of the systems in Croatia. The Decision Unit (DU) is the Ministry of Regional Development, Forestry and Water Management of the Republic of Croatia. The public agency "Hrvatske vode" ("Croatian Waters") serves as an Expert Unit for both systems.

The main activities of the Sava Commission in the field of flood management, pursuant to Article 13, Paragraph 1 of the FASRB are the comprehensive assessment of current flood management practices in the Parties and the preparation and elaboration of a draft protocol regulating flood protection in the Sava River Basin. The assessment activities have been performed by collecting the information related to hydrology, hydraulics and mapping practices in the Sava countries, the regulatory and institutional responsibilities as well as the flood protection structures. The main outputs of the assessment were expected to be the information on flood-prone areas in the Sava River Basin, the characteristics of flood events in the Sava River Basin, the responsibilities for flood management of the Parties, the flood protection structures and their state, the national flood prediction and warning practices as well as the review of strategic plans and regulatory acts. The collected GIS data assessment would be added to the Sava RBA Report in the form of a separate annex.

The draft protocol on flood protection has been prepared and distributed to all Parties with the aim to support the cooperation of all Parties in line with the EU Flood Directive, including joint activities in the preliminary flood risk assessment, the flood mapping, i.e. the preparation of the Flood Risk Management Plan for the Sava River Basin, the mutual assistance and the hydrologic information exchange, flood forecasting and warning system. The first draft of the Action Programme for Preparation of the Sava RB Flood Risk Management Plan has also been prepared.

Based on the Assessment of the Geographic Information System Capabilities of the FASRB Parties and the ISRBC Secretariat, the Geographic Information System Strategy for the Sava River Basin (Sava GIS Strategy), was finalized and adopted by the Sava Commission in 2008. The Strategy aims at establishing an effective and efficient (geo)information system and spatial data infrastructure that supports a wide range of water management planning functions and related activities in the framework of the Sava Commission, as well as to ensure that core geospatial information is available for use in multiple ways for the benefit of the ISRBC. The Strategy was prepared as a response to the requirements prescribed by the FASRB, but also takes into account the INSPIRE Directive for the establishment of an European spatial data infrastructure and WISE (Water Information System for Europe), as a wider initiative to modernize and streamline the collection and dissemination of information related to European water policy.

The Ministry of Environmental Protection, Physical Planning and Construction, as the main coordinator, works on the harmonization of spatial databases on the national level in Croatia. The Ministry coordinates and harmonizes spatial databases for all government sectors (Croatian Environmental Agency, State Geodetic Administration etc.).

3.3 *Status of Legislation and Institutional Framework*

The drafting and negotiation of new protocols to the FASRB (on flood protection, on emergency situations, on transboundary impacts and on sediment management) is an ongoing process.

In the Republic of Croatia, almost all the relevant documents for implementation of the FASRB were adopted. Among them the most important are the Water Management Strategy ("Official Gazette", No. 91/08), the National Flood Defence Plan ("Official Gazette", No. 8/97, 32/97, 43/98 and 93/99), the National Water Protection Plan ("Official Gazette", No. 8/99) and the River Transportation Development Strategy ("Official Gazette", No. 5/08).

The Ministry of Sea, Transport and Infrastructure and the Ministry of Regional Development, Forestry and Water Management are the institutions authorized to implement the FASRB in the Republic of Croatia.

3.4 Status of Cooperation

In accord with the FASRB, the cooperation with the Sava Commission and the Danube Commission (for navigation) is performed by means of mutual participation at sessions and expert group meetings of the Commissions, as well as at other events organized by any of the Commissions. As a legal basis for this cooperation, two memorandums of understanding on cooperation and coordination were signed between these two river Commissions: the Memorandum of Understanding on Cooperation and Coordination signed in Belgrade on June 5, 2008 and the Memorandum of Understanding on Cooperation signed in Budapest, on January 29, 2009.

The ICPDR, the Danube Commission and the Sava Commission guided jointly the implementation of the Joint Statement on Guiding Principles for the Development of Inland Navigation and Environmental Protection in the Danube River Basin.

The Sava Commission has been active in the preparatory phase of the Second Assessment of Transboundary Waters in the UNECE Region, whose draft was presented at the 5th Meeting of the Parties to the UNECE Water Convention which took place in fall 2009.

The Sava Comission has also been active in the Framework of the Monitoring and Assessment, Inland Water Transport and Standardization of Technical and Safety Requirements in Inland Navigation UNECE working groups.

Cooperation of the Sava Commission with the European Commission was based on the participation of the Sava Commission representatives in the working groups of DG TREN (Directorate-General Energy and Transport) and DG ENV (Directorate-General Environment).

The Sava Commission cooperate with other river navigation commissions like the Mosel Commission and the Central Commission for the Navigation on the Rhine, with other river commissions such as the Commissions for the protection of the Rivers Rhine, Elbe, and Oder, with regional organizations such as the Organization for Security and Co-operation in Europe (OSCE), the Regional Cooperation Council (RCC), the South East Europe Transport Observatory (SEETO), with financial organizations such as the World Bank (WB), the European Bank for Reconstruction and Development (EBRD), the European Investment Bank (EIB), the United States Agency for International Development (USAID) as well as with specialized organizations, associations and groups such as the World Meteorological Organization

(WMO), the Geographical Information Systems Forum (GIS Forum) and the World Association for Waterborne Transport Infrastructure (PIANC).

The Sava Commission is a member of the Global Water Partnership (GWP) and in this framework it cooperates with the Global Water Partnership Mediterranean (GWP-Med).

The Commission cooperates with the Ministries of the Republic of Croatia which are authorized for the implementation of the FASRB. Their representatives participated at the sessions and other events organized by the Sava Commission in relation to the preparation of the Sava River Basin Management Plan, the rehabilitation and development of the transport on the Sava River as well as the preparation and development of other projects relevant for the Sava River Basin (Ministry for Regional Development, Forestry and Water Management, Ministry for Environmental Protection, Physical Planning and Construction, Ministry for Foreign Affairs and European Integration, Ministry of Interior, Ministry of the Sea, Transport and Infrastructure).

The Sava Commission cooperates with environmental, non-governmental international, national organizations, local governmental, self-governmental bodies and relevant stakeholders aiming at ensuring the transparency of the FASRB implementation. Their representatives were invited to participate on sessions organized by the Sava Commissions in relation to the FASRB implementation monitoring, preparation of the Strategy for implementation of the FASRB and preparation of the protocols to the FASRB.

The Sava Commission organized or co-organized a number of consultation workshops, public presentations and other meetings with stakeholders. For example, meetings were held with members of the shipping industry and the representatives of the ports, as well as meetings with the local communities within the Sava Day celebration, with the general public to present the Draft Strategy for Implementation of the FASRB, workshops with the general public to present the Environmentally Sustainable Management and Maintenance of Inland Waterways, capacity building workshops discussing the Shared Groundwater Resources Management topics, meetings with the general public to present the Feasibility Study and Project Documentation for the Rehabilitation and Development of the Transport and Navigation on the Sava River Waterway, meetings with the general public presenting the Joint Statement on Guiding Principles for the Development of Inland Navigation and Environmental Protection in the Danube River Basin as well as consultation workshops discussing with the general public the Hydromorphological Issues in the Sava River Basin.

4 Conclusion

From the standpoint of the Republic of Croatia, it should be noted that the Republic of Croatia has clearly shown its support to the FASRB through the 2002 ratification and has been actively involved in all activities which are taking place within the Sava Commission.

In general, the implementation of the FASRB had a positive impact on improving the mutual contacts of the contracting parties, in the post-war era, giving them the opportunity to adopt specific knowledge and learn from the experiences of other countries. Through the cooperation in the FASRB frameworks and the positive effects achieved in the navigation sector, the Parties are now fully aware of the need for a greater level of cooperation in the field of water management. Over the four years of the Sava Commission's intensive activities, an even larger need has became apparent to consistently implement the EU *acquis* in the development of the river basin management plan for the Sava River as part of the international Danube River Basin. The FASRB initiated cooperation in the field of water management, with the aim to harmonize the water policy in the Sava River Basin with the water policy of the European Union. In this sense, it is visible that the Sava Commission, by means of the activities carried out by its Expert Groups, should pay much more attention to the requirements of the relevant EU directives.

The signature of the Memorandum of Understanding with the ICPDR and the Danube Commission (for navigation) as well as other relevant international institutions is an important motivation for the improvement of working of the Sava Commission.

Finally, with regards to navigation, important steps have been made in the establishment of an international navigation regime on the Sava River. It is also worth noting that, taking into consideration all hardships and obstacles, some progress has been made in the establishment of a sustainable water management and measurement policy for protection of waters from pollution and prevention or limitation of hazards and floods.

In terms of current activities on the implementation of joint projects of the FASRB Parties, it should be pointed out that all of them have been carried out in line with the adopted Agreement Implementation Strategy, whose goal is to encourage priority activities in the countries along the Sava River. In the case of the Republic of Croatia, these activities are implemented through the institutions responsible for water management, i.e. Ministry of Regional Development, Forestry and Water Management, Croatian Waters, Meteorological and Hydrological Service and others, which are involved into the work of Expert Groups of the Sava Commission.